Mathias Graf

Bayesian framework for probabilistic modelling of typhoon risks

AF061373

Mathias Graf

Bayesian framework for probabilistic modelling of typhoon risks

Südwestdeutscher Verlag für Hochschulschriften

Impressum / Imprint

Bibliografische Information der Deutschen Nationalbibliothek: Die Deutsche Nationalbibliothek verzeichnet diese Publikation in der Deutschen Nationalbibliografie; detaillierte bibliografische Daten sind im Internet über http://dnb.d-nb.de abrufbar.

Alle in diesem Buch genannten Marken und Produktnamen unterliegen warenzeichen-, marken- oder patentrechtlichem Schutz bzw. sind Warenzeichen oder eingetragene Warenzeichen der jeweiligen Inhaber. Die Wiedergabe von Marken, Produktnamen, Gebrauchsnamen, Handelsnamen, Warenbezeichnungen u.s.w. in diesem Werk berechtigt auch ohne besondere Kennzeichnung nicht zu der Annahme, dass solche Namen im Sinne der Warenzeichen- und Markenschutzgesetzgebung als frei zu betrachten wären und daher von jedermann benutzt werden dürften.

Bibliographic information published by the Deutsche Nationalbibliothek: The Deutsche Nationalbibliothek lists this publication in the Deutsche Nationalbibliografie; detailed bibliographic data are available in the Internet at http://dnb.d-nb.de.

Any brand names and product names mentioned in this book are subject to trademark, brand or patent protection and are trademarks or registered trademarks of their respective holders. The use of brand names, product names, common names, trade names, product descriptions etc. even without a particular marking in this works is in no way to be construed to mean that such names may be regarded as unrestricted in respect of trademark and brand protection legislation and could thus be used by anyone.

Coverbild / Cover image: www.ingimage.com

Verlag / Publisher:
Südwestdeutscher Verlag für Hochschulschriften
ist ein Imprint der / is a trademark of
AV Akademikerverlag GmbH & Co. KG
Heinrich-Böcking-Str. 6-8, 66121 Saarbrücken, Deutschland / Germany
Email: info@svh-verlag.de

Herstellung: siehe letzte Seite /
Printed at: see last page
ISBN: 978-3-8381-3651-6

Zugl. / Approved by: Zürich, ETH, Diss, 2012

Copyright © 2013 AV Akademikerverlag GmbH & Co. KG
Alle Rechte vorbehalten. / All rights reserved. Saarbrücken 2013

Acknowledgements

I would like to thank Professor Dr. Michael Havbro Faber and Professor Dr. Kazuyoshi Nishijima for their valuable supervision and support during my PhD studies. I am grateful for the many interesting discussions we had and their critical comments have always been an essential element in the completion of the present work.

I would also like to thank Professor Dr. Eleni Chatzi for her great support and for taking over the role as referee. My gratitude goes also to Professor Dr. Antoine Bommier, and to Professor Dr. Masahiro Matsui for acting as co-examiners and providing me with valuable comments. For proofreading my thesis I like to thank Dr. Harikrishna Narasimhan.

I would like to greatly acknowledge the support from Aon Benfield Japan in the joint research project, which was successfully completed during my PhD studies and which is the basis of this thesis. I would like to especially thank Sayaka Ochi for the good cooperation. Aon Benfield Japan provided historical damage observations which enabled to establish a data based vulnerability model to estimate portfolio loss distributions as described in Chapter 6. In order to fulfill the requirements of the confidentiality agreement between Aon Benfield Japan and ETH Zurich the damage data and all results which are obtained by using the damage data are made anonymous and censored in this thesis.

I also would like to thank Professor Takashi Maruyama for the cooperation in the study on climate change in Section 7.2. He provided the output of a climate model and a vulnerability model, which were used as a basis for this study.

Last but not least, I like to thank my parents for all their support during my studies.

Zürich, April 2012   Mathias Graf

Abstract

Risk assessment due to typhoon events is an important issue for decision makers in many different areas like the insurance markets or in governance. Towards this end a number of typhoon models have been developed so far. Whereas the development of improved models is one of the relevant issues to enhance optimal decision making, an equally or even more important issue is to develop a framework that can incorporate all relevant available information. Using all the available information enables a more precise risk assessment in two different ways: (1) For a risk assessment of an approaching typhoon event, actual information can be used to condition the model. This reduces the uncertainties for a specific risk analysis. (2) By using the information to update the parameters in the model, the overall modelling uncertainties can be reduced over time.

The objective of this thesis is to develop a Bayesian framework for the probabilistic modelling of typhoon risks. This framework includes not only the methodology to establish a typhoon model but also considers consistently the involved uncertainties and provides the mechanisms which enable the decision makers to take into account the available information during the process of decision making and thus facilitate conditioning the model and updating the parameters in the model over time.

The methodology employed in the thesis takes basis in Bayesian statistics and Bayesian probabilistic modelling, which provides the rationale to condition and to update the probabilistic model with additional data. As an example, the Bayesian framework is applied to the region of Japan and a typhoon risk model is established with the focus on the following features which facilitate the incorporation of the available information: typhoon events are modelled for the entire life of typhoons, i.e. from occurrence to dissipation; the effects of sea surface temperature (hereafter, SST) on the evolution of typhoon events are accounted for; seasonal differences in the probabilistic characteristics of the transition of typhoons are accounted for. The typhoon model established based on the proposed framework consists of two parts; a hazard model and a vulnerability model. The hazard model is composed of sub-models, describing all phases of the typhoon hazard process starting with the occurrence of typhoons over the spatial and temporal development of typhoons including landfall and possible filling and ending with the probabilistic characterization of extreme wind speeds at any location in Japan. The vulnerability model represents the probability distribution of the loss of individual exposures as a function of the wind speed. Based on the developed typhoon model, the consistently consideration of the involved uncertainties are investigated and a framework for updating the model is proposed.

The usefulness of the proposed Bayesian framework is demonstrated in three practical examples. The application to insurance portfolio risk analysis shows how the uncertainties and the correlation between the individual losses can be considered. The second application shows how the Bayesian framework can be used to assess the effect of a climate change by conditioning and updating the model with new available information and data. The third application demonstrates how the proposed Bayesian framework can be used for real-time decision making in the case of an approaching typhoon by conditioning the models with the new available information.

The proposed framework is implemented in several software tools, which facilitate the practical use of the proposed Bayesian framework. These software tools allow a user-friendly typhoon risk analysis and to condition and update the typhoon model.

The scientific contribution of this thesis is a better phenomenological insight on typhoon events in probabilistic terms and provides a first step to a full probabilistic treatment of typhoon risks. The societal benefit is to enhance decision making by providing a framework where the underlying uncertainties can be reduced by incorporating all available information. The implementation of the Bayesian framework in a software tool supports decision makers in practical applications.

Zusammenfassung

Risikoabschätzungen von Taifunereignissen ist eine wichtige Aufgabe für Entscheidungsträger in vielen verschiedenen Gebieten, wie zum Beispiel den Versicherungsmärkten oder in Regierungen. Für diesen Zweck wurden schon einige Taifun-Modelle entwickelt. Während die Entwicklung von verfeinerten Modellen wichtig für die Verbesserung der optimalen Entscheidungsfindung ist, ist es ähnlich wichtig oder sogar noch wichtiger, ein Framework zu entwickeln, welches es erlaubt, alle relevanten verfügbaren Informationen zu integrieren. Die Verwendung dieser Informationen ermöglicht eine präzisere Risikoabschätzung in zwei verschiedenen Situationen: (1) Während der Risikoabschätzung für einen sich nähernden Taifun können aktuelle Informationen verwendet werden, um das Modell zu konditionieren. Dies reduziert die Unsicherheit für die spezifische Risikoanalyse. (2) Die Verwendung aller verfügbaren Informationen zur Aktualisierung der Modelparameter reduziert die Modelunsicherheit über die Zeit hinweg.

Das Ziel dieser Dissertation ist es, ein Bayes'sches Framework für die probabilistische Modellierung von Taifunrisiken zu entwickeln. Dieses Framework beinhaltet nicht nur die Methodik ein Taifunmodell zu erstellen, sondern berücksichtigt konsistent die involvierten Unsicherheiten und bietet einen Mechanismus, welcher dem Entscheidungsträger ermöglicht, die verfügbaren Informationen während des Entscheidungsprozesses zu berücksichtigen um das Modells zu konditionieren und die Modellparameter über die Zeit hinweg zu aktualisieren.

Die Methodik, welche in dieser Dissertation angewandt wird, basiert auf der Bayes'schen Statistik und der Bayes'schen Wahrscheinlichkeitsmodellierung. Diese stellen die Grundprinzipien zur Verfügung, um das probabilistische Modell mit neuen Informationen und Daten zu konditionieren und zu aktualisieren. Das Bayes'sche Framework wird als Beispiel für die Region des Nordwest-Pazifiks angewendet. Das Taifunmodell wurde entwickelt mit Fokus auf den folgenden Eigenschaften, welche es ermöglichen, alle verfügbaren Informationen zu verwenden: Die gesamte Lebensdauer eines Taifunereignisses, von der Entstehung bis zur Auflösung, wird modelliert; der Effekt der Meeresoberflächentemperatur auf die Entwicklung der Taifune wird berücksichtigt; saisonale Unterschiede der probabilistischen Eigenschaften der Bewegung und Entwicklung der Taifune sind berücksichtigt. Das auf dem vorgeschlagenen Framework basierend entwickelte Taifun-Modell besteht aus zwei Komponenten; einem Gefahrenmodell und einem Schadensmodell. Das Gefahrenmodell besteht aus Sub-Modellen, welche alle Phasen eines Taifuns beschreiben, beginnend mit der Entstehung eines Taifuns, über die räumliche und zeitliche Entwicklung eines Taifuns, bis zur probabilistischen Charakterisierung extremer Windgeschwindigkeiten an beliebigen Orten in Japan. Das Schadensmodell repräsentiert die Wahrscheinlichkeitsdichtefunktion des Verlustes als eine Funktion der Windgeschwindigkeit. Basierend auf dem entwickelten Taifun-Modell wird die konsistente Berücksichtigung der involvierten Unsicherheiten untersucht und ein Framework für das Aktualisieren des Modells vorgeschlagen.

Die Nützlichkeit des vorgeschlagenen Bayes'schen Frameworks ist in drei praktischen Beispielen demonstriert. Die Anwendung des Frameworks für die Risikoanalyse eines Versicherungsportfolio zeigt, wie die Unsicherheiten und die Korrelationen zwischen den einzelnen Verlusten berücksichtigt werden können. Die zweite Anwendung zeigt, wie das Bayes'sche Framework verwendet werden kann, um den Effekt eines Klimawandels, durch Konditionieren und Aktualisieren des Models mit neu verfügbaren Informationen und Daten, abzuschätzen. Die dritte Anwendung demonstriert, wie das Bayes'sche Framework durch Konditionierung des Modells mit neuen Information verwendet werden kann um Echtzeit-Entscheidungsfindungen zu unterstützen für die Situation, dass ein neuer Taifun sich nähert.

Das vorgeschlagene Framework ist in einige Softwaretools implementiert, welche die praktische Benutzung des Bayes'schen Frameworks ermöglichen. Diese Softwaretools ermöglichen eine benutzerfreundliche Taifunrisikoanalyse und das Konditionieren und das Aktualisieren des Taifunmodells.

Der wissenschaftliche Beitrag dieser Dissertation ist, dass das Phänomen Taifun in Begriffen der Statistik besser verstanden wird und bietet einen ersten Schritt zu einer komplett probabilistischen Behandlung der Taifunrisiken. Der soziale Nutzen ist es, Entscheidungsfindungen zu verbessern durch die Zurverfügungstellung eines Frameworks, welches durch die Verwendung alle verfügbaren Informationen die Unsicherheiten verringert. Das Framework wurde in ein Softwaretool implementiert, um die Entscheidungsträger bei ihrer Arbeit zu unterstützen.

Table of Contents

ACKNOWLEDGEMENTS 1

ABSTRACT 3

ZUSAMMENFASSUNG 6

TABLE OF CONTENTS 9

1. INTRODUCTION 15
    1.1. BACKGROUND AND MOTIVATION 15
    1.2. AIM 17
    1.3. SCOPE 18
    1.4. STATE OF THE ART 21
    1.5. HYPOTHESIS 30
        1.5.1. Typhoon model 30
        1.5.2. Treatment of epistemic uncertainties in the typhoon model 32
        1.5.3. Updating the typhoon model 33
        1.5.4. Portfolio risk analysis 33
        1.5.5. Global warming 34
        1.5.6. Risk assessment of a approaching typhoon 34
    1.6. OUTLINE OF THE THESIS 34

2. TYPHOON MODEL 36
    2.1. MAIN FEATURES OF THE DEVELOPED TYPHOON MODEL 36
    2.2. APPLICATIONS OF THE DEVELOPED TYPHOON MODEL 37
    2.3. COMPONENTS OF THE TYPHOON MODEL 38
    2.4. UTILIZED DATASETS 40
    2.5. OCCURRENCE MODEL 44

    2.5.1. Definition of the initiation of typhoon events 45

    2.5.2. Probabilistic model for the occurrence of typhoons 46

  2.6.   TRANSITION MODEL 48

    2.6.1. Probabilistic model for the translation of typhoons 50

    2.6.2. Probabilistic model for the central pressure of typhoons 52

    2.6.3. Interpolation of the typhoon tracks 53

    2.6.4. Typhoon lysis 54

  2.7.   FILLING MODEL 55

  2.8.   PROBABILISTIC MODEL FOR THE RADIUS OF MAXIMUM WIND SPEED 56

  2.9.   WIND FIELD MODEL 57

    2.9.1. Model for pressure fields 57

    2.9.2. Model for wind field as a function of pressure field 58

    2.9.3. Calculation of maximum 10-minute sustained wind speeds 60

    2.9.4. Wind direction 62

  2.10.   SURFACE FRICTION MODEL 63

    2.10.1. Relation between the wind speeds at gradient height and at nominal height 63

    2.10.2. Relation between the wind direction at gradient height and at nominal height 65

    2.10.3. Roughness length 67

    2.10.4. Topography 72

  2.11.   SOFTWARE TOOL FOR CREATING A HAZARD EVENT SET USING THE HAZARD MODEL 75

3.   **VERIFICATION AND VALIDATION OF THE TYPHOON MODEL** 76

  3.1.   OCCURRENCE OF TYPHOONS 76

  3.2.   TRANSITION OF TYPHOONS 77

    3.2.1. Extrapolation of the transition model to the future 83

    3.2.2. Seasonal differences in the transition model 84

  3.3.   WINDS INDUCED BY TYPHOONS 87

  3.4.   WIND HAZARD MAP 90

4. TREATMENT OF EPISTEMIC UNCERTAINTIES IN THE TYPHOON MODEL 92

- 4.1. INTRODUCTION ... 93
- 4.2. BACKGROUND ... 94
- 4.3. STATUS IN PRACTICE AND STATE-OF-THE-ART ... 95
- 4.4. CHALLENGING ISSUES ... 96
- 4.5. GENERAL FRAMEWORK FOR UNCERTAINTY TREATMENT ... 97
- 4.6. REFERENCE TYPHOON MODEL ... 101
- 4.7. OVERVIEW OF THE TYPHOON MODEL ... 101
- 4.8. TRANSITION MODEL ... 102
- 4.9. ALTERNATIVE MODELS ... 105
  - 4.9.1. Discretization in space and time ... 106
  - 4.9.2. Functional form of transition ... 106
  - 4.9.3. Functional form of intensity ... 107
  - 4.9.4. Data sets ... 107
- 4.10. OVERVIEW OF THE ALTERNATIVE MODELS ... 107
- 4.11. VARIABILITY OF HAZARD ASSESSMENT BETWEEN ALTERNATIVE MODELS ... 108
  - 4.11.1. Variation of the statistics on typhoon transition ... 108

5. UPDATING OF THE TYPHOON MODEL ... 114

- 5.1. BACKGROUND ... 115
- 5.2. INTRODUCTION ... 116
- 5.3. PROBLEM SETTING ... 117
- 5.4. PROPOSED APPROACH ... 119
- 5.5. EXAMPLES ... 123
  - 5.5.1. Proposed approach vs. standard approach for updating fragility models with data ... 124
  - 5.5.2. Updating of fragility models with the presence of common uncertainties ... 129

      5.5.3. Updating of fragility model using hazard intensities measured at meteorological stations 133

    5.6.   MODEL BUILDER SOFTWARE TOOL 134

6. APPLICATION: PORTFOLIO RISK ANALYSIS 136

    6.1.   VULNERABILITY MODEL 137

      6.1.1. Treatment of uncertainties 138

      6.1.2. Exposure data and loss data available 139

      6.1.3. The flow of the development of the vulnerability model 143

      6.1.4. Conditional probability distribution of loss ratio given occurrence of loss 145

      6.1.5. Probability of occurrence of loss 147

      6.1.6. Unconditional probability distribution of loss ratio 151

      6.1.7. Dependency of the random variables $\varepsilon_{s,c}$ 153

    6.2.   INSURED LOSSES CAUSED BY HISTORICAL TYPHOONS 154

    6.3.   ASSESSMENT OF TYPHOON RISKS 155

      6.3.1. Mathematical formulation for assessment 155

      6.3.2. Typhoon event sets 164

      6.3.3. Disaggregation of portfolios 165

      6.3.4. Program components 169

    6.4.   *TRAST*: SOFTWARE TOOL FOR PORTFOLIO RISK 171

      6.4.1. Defining the policy conditions 172

      6.4.2. Defining the analysis conditions 174

      6.4.3. Displaying analysis results 176

7. APPLICATION: GLOBAL WARMING RISK ASSESSMENT 182

    7.1.   ADAPTION OF TYPHOON RISK MODELLING TO CLIMATE CHANGES 183

      7.1.1. Objectives of the study 184

      7.1.2. Approach adopted in the study 185

      7.1.3. Validation of the adopted hazard model for wind loads assessment 186

7.1.4. Assessment of the reliability of structures 186

7.1.5. Example 188

7.1.6. Summary 191

7.2. A PRELIMINARY IMPACT ASSESSMENT OF TYPHOON WIND RISK OF RESIDENTIAL BUILDINGS IN JAPAN UNDER FUTURE CLIMATE CHANGE 192

7.2.1. Introduction 192

7.2.2. Approach 194

7.2.3. Results 200

7.2.4. Discussion 206

7.2.5. Surface roughness and development of the society 208

7.2.6. Acknowledgements 209

8. APPLICATION: RISK ASSESSMENT OF A APPROACHING TYPHOON AND REAL-TIME DECISION MAKING 210

8.1. *TRAST*: RISK ASSESSMENT OF A APPROACHING TYPHOON 211

8.1.1. Scenario-based simulation 211

8.1.2. Conditional simulation 211

8.2. REAL TIME DECISION MAKING 213

8.3. CHARACTERIZATION OF THE PROBLEM 215

8.4. DECISION FRAMEWORK 216

8.4.1. Conditional probability representations 216

8.4.2. Decision optimization 218

8.4.3. One-time decision making 220

8.4.4. Sequential decision making 221

8.5. EXAMPLE 222

8.5.1. Problem setting 222

8.5.2. Typhoon model 223

8.5.3. Postulated consequence model 223

8.5.4. Other conditions 224

		8.5.5. Algorithm	225
		8.5.6. Results	230
		8.5.7. Discussions	231
9.	**CONCLUSIONS AND OUTLOOK**		**234**
	9.1.	SUMMARY	234
	9.2.	CONCLUSIONS	236
		9.2.1. Typhoon model	237
		9.2.2. Treatment of epistemic uncertainties in the typhoon model	238
		9.2.3. Updating the typhoon model	241
		9.2.4. Portfolio risk analysis	242
		9.2.5. Global warming	243
		9.2.6. Risk assessment of a approaching typhoon	245
	9.3.	SCIENTIFIC ACHIEVEMENTS	246
	9.4.	OUTLOOK	248
10.	**APPENDIX**		**250**
	10.1.	APPENDIX A - VERIFICATION OF THE TRANSITION MODEL	250
	10.2.	APPENDIX B - PARAMETERS FOR ESTIMATING THE TOPOGRAPHY FACTOR	254
	10.3.	APPENDIX C - EQUIVALENCE OF THESE BAYESIAN PROBABILISTIC NETWORKS IN SECTION 5.5	256
	10.4.	APPENDIX D - *TRAST* VISUALIZATIONS	259
11.	**NOMENCLATURE**		**260**
12.	**REFERENCES**		**264**
	LIST OF FIGURES		**283**
	LIST OF TABLES		**290**

1. Introduction

1.1. Background and motivation

Efficient processing of information and consistent modeling of physical phenomena are important requisites for rational risk management of natural hazards. Any tool for the management of risks due to natural hazards should be developed to provide a sound basis for optimizing decision making with the purpose of reducing risks, and at the same time facilitate the incorporation of all relevant information and best knowledge available.

Consistent and precise loss estimation due to typhoon damages is an important issue for decision makers, for example in the insurance markets. Typhoon models have been developed to estimate portfolio losses. The results from the models help the decision makers by determining insurance premiums for portfolios. So far, several models have been proposed and implemented into software tools, whereby the parameters which the decision makers are required to input are the information on the portfolio, e.g., locations and values of buildings in the portfolio and the insurance policy etc.. All the other parameters related to the phenomenological characteristics of typhoons and the meteorological environment surrounding typhoons, which are necessary to estimate the portfolio losses, are fixed and cannot be changed by the decision maker as a user of the software. In many cases some additional information or data or strong belief, which the decision makers may possess, is available; there seems no way to reflect this in the process of loss estimation in the presently available software tools.

Whereas the development of more sophisticated and improved models is a key to keep advantages in the insurance markets, this is not the only way and possibly not the most relevant way. In order to explain the idea behind this thesis, three terminologies are differentiated: knowledge, data and information and the corresponding terminologies: modeling, updating and conditioning, see Figure 1.1. Seen in the light of this differentiation, most available software tools have been aimed to develop the models by implementing more scientific and professional knowledge. The philosophy behind developing the presently available software is to take into account as much professionally reliable knowledge as possible into the model. Once the model is fixed, there is no need to change it in daily use. This thesis, explores this direction to some extent, but this is not the main target. Instead, a Bayesian framework is established within which the data and the information which the decision makers possess can be implemented into the model runningly as information becomes available, namely, the model is updated by data and conditioned by information. Here, information refers to factual information or a belief which is available to a decision maker in the decision making process and data refers to an organized set of information in such a way that the data can be utilized for updating the parameters in the model. For instance, sea surface temperature is information which may affect the intensity of typhoons, while the accumulation of information on the relationship between sea surface temperature and the intensity of typhoon can be used to update the model. In a decision process, if such information is available, the decision maker should implement it into the model (conditioning), because the conditioned model gives an estimation of portfolio loss with a smaller variance and thereby enhances decision making. The parameters in the model should be modified with data (updating), because the updated model gives more reliable estimations. Apart from the professional knowledge necessary to construct a basic model, such information and data are readily available and accessible even to decision makers. The full utilization of data and information which becomes available over time is another key to ensure market advantage and competiveness.

1. Introduction

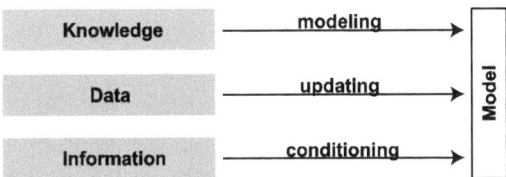

Figure 1.1: Integration of knowledge, data and information.

To ensure the consistent use of the available information and data it is essential to consider the uncertainties involved in the modelling of the phenomena. This type of variability is generally understood as epistemic uncertainty, which arises from the lack of sufficient data and/or knowledge. Note that in contrast to the epistemic uncertainty, randomness in nature is called aleatory uncertainty. The general treatment of both types of uncertainty in risk assessment and formal decision analysis has been, since decades, an issue of attention in civil engineering and other fields.

However, in practice the epistemic uncertainty is often ignored, mixed up with aleatory uncertainty otherwise treated in ad hoc manners, which can lead to erroneous assessment of risks. Such examples are investigated in detail in Nishijima et al. (2008b).
Considering consistently the involved uncertainties allow a more precise and realistic assessment of the typhoon risk.

1.2. Aim

The dissertation aims to develop a Bayesian framework for the probabilistic modelling of typhoon risk for the quantitative and rational risk management. The proposed framework is designed for decision-makers responsible for insurance portfolios, for the assets of large areas such as cities or the safety of personnel. The framework is generic in the sense that it is formulated in terms of observable indicators and can thus be easily implemented for the characteristics of a specific region. The framework is applied as an example for the region of the North West Pacific and is calibrated and verified for the Japanese Islands. The aim of the Bayesian framework is to deliver a first step towards a full probabilistic treatment of typhoon risk analysis, which considers all involved uncertainties and is able to update and condition the typhoon model with new available data and information. The framework includes:

- a probabilistic typhoon model which satisfies the requirements for the proposed framework.
- the consideration of the uncertainties involved in the modelling of the typhoons and in the portfolio risk analysis.
- the means to update and condition the typhoon model.

The application of the framework is illustrated in three examples which show how the framework can be used for insurance portfolio risk analysis, for an assessment considering global warming and for real time decision making in the case of an approaching typhoon.

1.3. Scope

The objective of the present thesis is to develop a Bayesian framework for probabilistic modelling of typhoon risks. The framework includes not only a probabilistic typhoon model but also the mechanisms which enable the decision makers to reflect the information and data during the process of loss estimation. Most of the efforts in this thesis are devoted to establish a typhoon model rather than towards detailed modeling of the damage process of buildings due to typhoons.

Probabilistic modelling of natural hazard events generally aims at describing the probabilistic characteristics of the underlying physical phenomena associated with the events. The degree of detail and requirements to the probabilistic models should be determined in accordance with their applications, i.e. the type of risk management and more generally decision situations. Seen in this light, it can be said that most of the existing typhoon models are developed primarily for assessing wind hazards for the purpose of facilitating structural design in regard to wind loads and estimating individual exposure as well as portfolio losses. Whereas these are some of the most relevant and successful applications of the typhoon models, there are other relevant applications where the typhoon model can potentially be useful. The typhoon model developed during this thesis is developed with the scope of applying the model for a broader range of decision situations. Such decision situations include: real-time decision making for the evacuation of people and shut-down of engineered facilities in the face of emerging typhoon events; adaptation of building codes in regard to wind loads to the possible increase of wind hazards that might be caused by global climate change.

Taking basis in the state-of-the-art research work on the modelling of typhoon events, the presented typhoon model is developed to facilitate the incorporation of all available information. The focus is on the following features: (1) typhoon events are modelled for the entire life of typhoons, i.e. from occurrence to dissipation; (2) the effects of sea surface temperature (hereafter, SST) on the evolution of typhoon events are accounted for; (3) seasonal differences in the probabilistic characteristics of the transition of typhoons are accounted for. The developed typhoon model can, in principle, represent the wind hazards due to typhoons in the North West Pacific region, whereas the developed typhoon model is verified and validated primarily for the area of the Japanese islands.

The established framework is general, but the region of the North West Pacific is selected as an example. Three main features are included in the framework: (1) estimation of annual average loss/probable maximum loss;

(2) estimation of loss of any given portfolio when a typhoon event has initiated and is approaching the considered region; (3) updating of the models with all the data available after one or more typhoon events have occurred.

In the first feature, the framework facilitates the estimation of a portfolio loss when a decision maker determines the insurance premium year to year. During the process, if the decision maker possesses the information which conditions the model, e.g., the prediction of sea surface temperature in a year, this information can be implemented.

In the second feature, since the typhoon has already initiated, some information should be readily available, e.g., the track of the typhoon, the central pressure of the typhoon or meteorological environment surrounding the typhoon. This information conditions the model, enabling a loss estimation with a smaller uncertainty. This feature is useful, for instance, when a financial department in a company prepares for the post-disaster financing or when the decision maker makes an additional reinsurance contract to avoid a catastrophic insurance payment.

The third feature serves to update the typhoon model after a certain period of time for example at the end of a year, when all the information is organized as data. So, over time, the model will better represent the underlying physical phenomena.

The use of the framework is shown in three applications. The first application shows how this framework can be used to estimate portfolio losses due to typhoons with the consideration of the involved uncertainties and is implemented into a software decision support tool. The second application investigates how the framework can be used to assess the effects of global warming by updating the typhoon model with new available data. The third application demonstrates how the framework can be used for real time decision making in the case of an approaching typhoon by conditioning the typhoon model with new available information.

The proposed framework is implemented in three software tools. (1) a model builder software tool is developed, which enables to establish a typhoon model using all available information. (2) a hazard event set creator software tool is developed, which establishes a data base of stochastic typhoon events using the typhoon model to simulate typhoon events and (3) the software tool *TRAST* (Typhoon Risk Analysis Software Tool), which provides a user friendly graphical interface to perform a risk analysis for a insurance portfolio based on the hazard events set data base in combination with a vulnerability model.

1.4. State of the art

In the risk management and loss estimation for natural hazards, the physical modeling of the hazard event plays a key role. Since strong natural hazards are rare events, only a limited amount of recorded data is available at specific locations where measurement stations are located. To estimate the risk due to natural hazards the temporal and spatial distribution of the hazard index (e.g. for tropical cyclones the wind speed) is needed. To use the available data in a most efficient way, physical and statistical models are established and fitted to the data to represent the temporal and spatial distribution of the hazard index. A literature review of the development of natural hazard loss model can be found, for example, in (Walker, 1997) and specific for typhoons in (Nishijima and Faber, 2007).

The first frameworks for natural hazard risk modelling have been proposed by Friedman in the early 1970's for different natural hazards (Friedman, 1975). Typhoon/hurricane models which were based on Monte Carlo simulations were originally introduced in the field of offshore engineering (see e.g. (Russell, 1971)).

Over the last years the standard methodology for the probabilistic modeling of typhoon events has been established. Presently, typhoon models developed based on the methodology are widely utilized in the assessment of typhoon hazards and risks. The relevant practical applications include the identification of design wind loads in performance based structural design and portfolio loss analysis in the insurance industry, see e.g. (Matsui et al., 2002; Yasui et al., 2002; Emanuel, 2006a; Vickery et al., 2006a; Lee and Rosowsky, 2007; Rumpf et al., 2007). For example, the Architectural Institute of Japan proposes a design wind speed map for the design of structures subjected to strong wind in Japan (Architectural Institute of Japan, 2004). This is established on the basis of typhoon simulation as proposed by (Matsui et al., 2002). The HAZUS-MH Hurricane model, which also takes basis in hurricane simulation, is used for the prediction of possible losses of buildings subject to hurricane, see e.g. (Vickery et al., 2006a). The approaches described in the literature mentioned above correspond to a standard methodology which is comprised of the following model components:

- Occurrence model
- Transition model
- Wind field model
- Surface friction model
- Vulnerability model

Other literature on typhoon simulations suggest a similar methodology, see e.g. (Katsuchi and Yamada, 2005). Differences in the approaches described in the literature exist mainly in the development of each individual model component rather than in the general methodology. In the private industry several commercial hurricane/typhoon models are developed, for example, by Risk Management Solutions (RMS), EQECAT or Applied Insurance Research (AIR), for which no detailed information is publicly available.

Many typhoon/hurricane models have been developed for different hurricane prone regions e.g. Northern Atlantic, Northwest Pacific, Australian region etc. and in several different research fields e.g. in civil engineering, offshore engineering, wind engineering, atmospheric science, financial and insurance sectors. Depending on their requirements, research fields have different focus areas that define which components of the typhoon model are more important for their applications; these are therefore further enhanced.

The civil engineering community is interested in the wind load due to strong winds induced by typhoons on the structures. Since the majority of the structures are located on land, the greatest uncertainty in the estimation of strong winds due to typhoons comes from the different properties of the ground (i.e. land use) and the presence of escarpments, hills and ridges which act as obstacles to the boundary layer and accelerate the wind speed at surface level. Therefore more efforts are devoted to develop sophisticated surface friction models.

As a special field of civil engineering, offshore engineering is concerned about structures on the ocean and therefore on the wind speeds due to typhoons on the sea. Since the surface of the water has a quite low friction and the surroundings of the considered location are homogenous (i.e. water) and have no obstacles (e.g. hills) the offshore engineers do not have to develop a detailed surface friction model. They focus on enhancing the transition and the wind field models. In addition to the estimation of the strong winds, offshore engineers are also interested in the estimation of the wave heights induced by typhoons.

During 1950-1977, typhoon/hurricane wave models based on wave height and period were developed (e.g. (Bretschneider, 1959; Ross, 1976)). Young (1988) proposed a parametric hurricane wave model, which uses the radius of the maximum wind speed, the maximum surface wind and the transition speed of a typhoon as input parameters.
The wind engineers are more focused on the detailed wind speed on surface level and the influence of the wind load on structures. The common practice is to perform a Computational Fluid Dynamics (CFD) analysis of the considered structure and the surrounding area to analyze the wind flow and the wind load on the structure in detail (e.g. (Tomokiyo and Maeda, 2004; Kwok and Hitchcock, 2009)).

To define the boundary conditions for a CFD simulation, reanalyzed historical typhoon data from metrological measurement stations can be used. For the case that a typhoon is approaching and no measurement data is available or for estimating the probability distribution of the wind speeds for a selected location the relevant parameters from a hurricane/typhoon model are used to set the boundary conditions of a CFD simulation (e. g. (Ishihara et al., 2005)).

The atmospheric scientists focus their attention on weather forecasting and the prediction of typhoon development. For the prediction of the hurricane intensity, several models have been proposed; for example, the coupled hurricane intensity prediction scheme (CHIPS) (Emanuel, 1988) and statistical hurricane intensity prediction scheme (Ships) (DeMaria and Kaplan, 1994; DeMaria and Kaplan, 1999; DeMaria et al., 2005; DeMaria, 2009) analog for the region of the north west pacific the typhoon intensity prediction scheme (TIPS) (Fitzpatrick, 1997), which is extended to the statistical typhoon intensity prediction scheme (STIPS) (Knaff et al., 2005).

The climate and environmental science community investigates the effect of climate change on the intensity of tropical cyclones. Previous studies have indicated that global warming would increase the intensity of future tropical cyclones with a possibility of a decrease in numbers of relatively weak tropical cyclones and an increase of intense tropical cyclones, see (Emanuel, 1987; Knutson and Tuleya, 2004; Emanuel, 2005; Solomon et al., 2007; Knutson et al., 2010). Whereas the results from such previous studies depend largely on the employed climate models as well as the assumed scenarios on the emission of greenhouse gases e.g. CO_2, thus subject to significant uncertainties, it is likely that tropical cyclones will bring more often stronger wind to structures and consequently result in more severe consequences for societies. Yet the magnitude of such consequences and the effectiveness of implementing new policies on the design and maintenance of structures in the face of global warming are not obvious.

The financial and insurance sectors are interested in the financial losses due to hurricanes/typhoons. Therefore the focus here lies on the development of vulnerability models to represent the loss due to hurricanes/typhoons.

This thesis focuses on the wind speed as the hazard index which is used for the risk assessment. Rainfall during typhoon events is also a relevant factor which may influence the losses to buildings. Observations on the precipitation distribution during typhoon events have been made by e.g. (Frank, 1985; Atallah and Bosart, 2003). A quantitative modeling of rainfall during typhoon events is quite complicated, and only very simple models have been proposed. In the HAZUS-MH Hurricane model (Vickery et al., 2006a), the precipitation at any given location during a typhoon event is modeled, based on statistical analysis, as a function of the radius of maximum wind speed and the distance from the center of the typhoon.

The state of the art of the individual components of the typhoon models (occurrence, transition, wind field, surface friction and vulnerability model) are described below.

Occurrence model

Many existing typhoon models generate the simulated typhoons at the coastline (e.g. (Shapiro, 1983; Fujii and Mitsuta, 1986)), or at "simulation circles" around the site of interest(e.g. (Tryggvason et al., 1976; Georgiou et al., 1983; Vickery and Twisdale, 1995a)). Since these studies are mostly concerned about the wind speed at each individual site, the spatial dependency of wind speeds at different sites cannot be considered appropriately.

Another approach is to simulate the entire life of a typhoon, from the occurrence to the dissipation of a typhoon. Recent studies proposing typhoon models, which start the simulation of the typhoons on the ocean and then continue with the simulation of the transition and development of a typhoon till the dissolving of the typhoon. (e.g. (Yasui et al., 2002; Katsuchi and Yamada, 2005; Emanuel, 2006b). This approach enables the integration of additional parameters (e.g. the SST) which affect the development of a typhoon over time. Yonekura and Hall (2011) propose an occurrence model which also considers influence by El Nino–Southern Oscillation (ENSO).

Transition model

The common practice for the modeling of the transition of a typhoon is to perform a Monte Carlo simulation which takes basis in the nonhomogenous Markov process (see e.g. (Vickery et al., 2000; Yasui et al., 2002; Katsuchi and Yamada, 2005)). The translation speed, translation direction and intensity of a typhoon are simulated conditioned on the previous state of the relevant parameters which characterize the state of a typhoon. Hall and Jewson (2007) follow the same methodology but they propose a modeling scheme for transition modeling, which considers all tropical cyclones with different weights as a function of distance, instead of discretization.

If a typhoon makes a landfall, a filling model considers the decay of the intensity of a typhoon, which may be represented in terms of the change of the central pressure. Based on statistical analyses Batt (1980) proposes to model the change of the central pressure as a function of the time after landfall. Another approach is to model the decay of the intensity of a typhoon after landfall as a function of the time after landfall and the angle between the typhoon direction and the coastline at the point of landfall (Vickery and Twisdale, 1995b; Fujii, 1998).

Wind field model

In previous studies two different approaches can be found to model the spatial distribution of the wind speed as a function of the parameters describing the state of a typhoon obtained from the transition model. The first approach models the wind field without considering the pressure field; here parameters which describe the rate of decay of wind speed with respect to the distance from the center of typhoon are used, see e.g. (Miller, 1967; Brand et al., 1977).

The second approach models the wind speed based on the pressure field. The pressure field is normally modeled in accordance with (Schloemer, 1954b), or by assuming the extended model by (Holland, 1980). To obtain the wind speed the Navier-Stokes equations or variants of the Navier-Stokes equations under some assumptions are solved, e.g. (Yoshizumi, 1968; Shapiro, 1983; Meng et al., 1995b). This obtained wind speed corresponds to the wind speed at free atmosphere. To take the surface roughness into account, a surface friction model has to be established to represent the relationship between the wind speed at free atmosphere and the wind speed at surface level at the selected location.

The validity of the wind field model for the second approach is investigated e.g. in (Fujii, 1998; Nishijima et al., 2004), by comparing the reproduced wind field using the wind field model and the observed wind speeds at meteorological stations.

Surface friction model

If the wind field is modeled according to the second approach described above, the obtained wind speed corresponds to the wind speed at free atmosphere. The surface friction model describes the relation between the wind speed at free atmosphere and the wind speed at surface level by considering the surface friction of the selected location. In the literature, e.g. (AS1170.2, 1989; Wieringa, 1993; Simiu and Scanlan, 1996; ASCE7-98, 2000), the surface friction is characterized by the land use as well as the density and height of buildings. The affection of the wind speed is described by the surface roughness in the surface friction model. In the literature there are mainly two approaches for the estimation of the surface roughness. An overview can be found at (Grimmond and Oke, 1999).

The first approach is to estimate the surface friction using measurements of the gust ratio as proposed by (Davenport, 1961). This approach was further investigated by several authors see e.g. (Wieringa, 1976; Ashcroft, 1994). This approach enables to estimate the surface friction only for locations where wind speed measurements are available. For locations where no wind speed measurements are available, Wieringa (1986) proposed an interpolation scheme.

The second approach is to estimate the roughness length using the aerodynamic parameters of the surface morphometry. This approach was proposed by (Lettau, 1969) and is used e.g. by (Kondo and Yamazawa, 1986; Bartheelmie et al., 1993). The advantage of this approach is that it is possible to estimate the roughness length at any location where information about the surface morphometry (e.g. Land use data) is available.

Recent studies, e.g. (Itoi et al., 2005), suggest to use GIS information together with numerical simulation techniques to obtain a better surface friction representation, taking into account directional aspects.
The wind speed at surface level is also dependent on the topography of the surrounding area at the considered location. Escarpments, hills and ridges act as obstacles to the boundary layer and accelerate the wind speed at surface level. An overview of how the topographic effect is considered in the different codes can be found in (Ngo and Letchford, 2008).

Vulnerability model
The vulnerability model describes the relation between a hazard index (in case of typhoon normally the wind speed) and the damage or the loss associated with a building or a insurance portfolio composed of many buildings.

Many vulnerability models have been developed in the insurance sector. However, most of them are unpublished because they are based on confidential client data. Among the published literature, a general overview is given in (Klugman et al., 2004), an engineering approach is proposed in HAZUS-MH Hurricane model (Vickery et al., 2006b). Pinelli (2004) describes how the damage of buildings due to strong winds caused by hurricanes can be modeled. Watson and Johnson (2004) propose an approach on how different vulnerability models can be combined to assess the variability of different models.

1.5. Hypothesis

The hypothesis is that a Bayesian framework for the probabilistic modelling of typhoon risk together with phenomenological models of the typhoon event as well as associated damage/losses to the built environment can be developed such as to accommodate the means to condition the models with new available information and update the models with new available data and such as to accommodate the consistently consideration and treatment of the different types of uncertainty associated with the random processes and the modeling in general. The purpose of the Bayesian framework is to support decision making for different fields of the society, e.g. industry, public authorities and individuals.

In this thesis the components of the Bayesian framework are subsequently proposed, tested and verified in terms of comparing the model results with observable characteristics. The examples thereafter serve to demonstrate the usefulness of the framework and of the developed tools.

1.5.1. Typhoon model

The first step for the proposed Bayesian framework for the probabilistic modelling of typhoon risk is to establish a probabilistic typhoon model. This model has to be developed in such a way that it satisfies the requirements for the proposed Bayesian framework. The involved uncertainties have to be consistently considered and treated and the framework has to provide the feasibility to update the typhoon model with new available data and to condition the typhoon model with new available information.

Whenever possible, the typhoon model is established on existing state of the art research and whenever necessary the work of previous researchers is modified or new components are developed so that the typhoon model satisfies the requirements for the proposed framework. Sub-models have to be formulated for all phases of the typhoon hazard process starting with the occurrence of typhoons, over the spatial and temporal development of typhoons including landfall and possible filling and ending with the probabilistic characterization of extreme wind speeds at any location in Japan.

These results together with historical damage observations made available by Aon Benfield Japan facilitated the establishment of the portfolio loss distribution. Emphasis has been given to the consistent treatment of uncertainties facilitating that the contributions to the uncertainty associated with the total losses from each sub-model may be assessed.

The developed model facilitates risk updating such that the losses can be estimated probabilistically in the event of an evolving typhoon as a function of the available information regarding location, central pressure, direction and velocity.

1. Introduction

For the probabilistic typhoon model the following components are established:

- A new occurrence model is developed, which represents the occurrence of a typhoon as a function of the location, the season (month) and the SST.
- For the transition model the proposed approach from (Vickery et al., 2000) is followed, but the model is adapted for the region of the North West Pacific and modified so that the development of a typhoon is also a function of the season (month) and the SST.
- The wind field model is established as proposed by (Georgiou et al., 1983).
- The surface friction model is developed as proposed by (Meng et al., 1997). A new scheme for the estimation of the roughness length is developed which combines the two approaches described in the state of the art.
- A model to estimate the portfolio losses is developed containing a new vulnerability model which considers the epistemic uncertainties.

The hypothesis is that this typhoon model can be developed and be implemented in a hazard event set builder software tool, which facilitates to automatically create a stochastic event set using the described typhoon model. This typhoon model has to provide the feasibility of conditioning with new available information.

1.5.2. Treatment of epistemic uncertainties in the typhoon model

The proposed Bayesian framework has to be developed in such a way that the different types of uncertainty associated with the random processes and the modeling in general are consistently considered and treated. The epistemic uncertainties in the typhoon model due to the modelling of the phenomena have to be quantified for each sub model and the epistemic uncertainties due to the model selection and the assumptions in the typhoon model have to be considered.

1.5.3. Updating the typhoon model

The proposed Bayesian framework has to provide the means to update the typhoon model with new available data with all the data available after one or more typhoon events have occurred.
The hypothesis is that the Bayesian framework can be developed so that it facilitates the updating of the models and that a model builder software tool can be created which automatically updates the typhoon model by establishing the typhoon model using as input all the available information.

1.5.4. Portfolio risk analysis

The application of the Bayesian framework to the estimation of insurance portfolio risk analysis demonstrates the usefulness of the framework. The hypothesis is that this example can show how the Bayesian framework can be used to estimate the risk insurance portfolios due to strong winds induced by typhoons and to demonstrate how the uncertainties can be considered.

A model to estimate the portfolio losses has to be developed which considers the epistemic uncertainties and the correlation between the individual building losses. The portfolio loss model is implemented in Typhoon Risk Analysis Software Tool (*TRAST*) which provides a user friendly graphical interface to perform a portfolio risk analysis. *TRAST* uses for the risk analysis a database of a stochastic event sets, which was created using the hazard event set builder software tool.

1.5.5. Global warming

The hypothesis is that the Bayesian framework can be used to investigate the effects of climate change. Therefore two studies are performed. The first study shows how the typhoon model can be conditioned by an increased SST and how the change of structural reliability considering the effect of the increased SST on tropical cyclone activity can be assessed. The second study shows how the typhoon model can be updated by using the data obtained from the mesoscale meteorological model JMA-NHM. This in turn is used to estimate the change in the wind risk of residential buildings in Japan under a future climate.

1.5.6. Risk assessment of a approaching typhoon

The hypothesis is that the Bayesian framework can be used for real time decision making in the case of an approaching typhoon by conditioning the typhoon model with new available information. The function to perform a risk analysis for a approaching typhoon has to be implemented in *TRAST* and a framework for real-time decision analysis is provided.

1.6. Outline of the thesis

The thesis is structured as follows: Chapter 2 describes the developed typhoon model in detail. Chapter 3 shows the verifications and the validations of the developed typhoon model. Chapter 4 explains how the uncertainties of the typhoon model are treated and proposes a framework for the incorporation of the uncertainties due to the model selection in the risk assessment and the decision analysis. Chapter 5 explains how a hazard model can be updated by new information in a most efficient way. Further, the developed model builder software tool is presented which allows to establish a typhoon model containing all available information. Chapters 6 to 8 present three different applications of the developed typhoon model. In the first application described in Chapter 6, it is shown how the typhoon model can be used for a portfolio risk analysis. The proposed procedure is implemented in the software tool *TRAST* which provides a user friendly graphical interface to perform a portfolio risk analysis. Chapter 7 shows how the developed typhoon model can be used to perform studies to assess possible effects of global warming on the risk assessment. Chapter 8 illustrates how the developed typhoon model can be used for real time decision making in the case when a new typhoon is approaching. Chapter 9 discusses the conclusions of the thesis and identifies future tasks in the outlook.

2. Typhoon model

2.1. Main features of the developed typhoon model

The components of the typhoon model are developed in such a way that the typhoon model satisfies the requirements for the proposed Bayesian framework. The components of the typhoon model facilitate the consideration of the uncertainties and the typhoon model can be conditioned and updated with new available information and data.

Taking basis in the state-of-the-art research works on the modelling of typhoon events, the presented typhoon model is developed with the focus on the following features: (1) typhoon events are modelled for the entire life of typhoons, i.e. from occurrence to dissipation; (2) the effects of sea surface temperature (hereafter, SST) on the evolution of typhoon events are accounted for; (3) seasonal differences of the probabilistic characteristics of the transition of typhoons are accounted for. The developed typhoon model can, in principle, represent the wind hazards due to typhoons in the northwest Pacific region, whereas the developed typhoon model is verified and validated primarily for the area of the Japanese islands.

Within the present framework for risk management the entire lives of typhoons are modeled. An advantage of this approach is that it enables conditional loss estimation given a typhoon has developed, since the information on the current state of typhoon can be used to simulate the rest life of the typhoon. This is especially useful in practical situations where decision makers in e.g. local authorities prepare for precautious actions as well as post-disaster actions.

2.2. Applications of the developed typhoon model

One of the main applications of the developed typhoon model is to estimate the statistics of insured portfolio losses in the insurance industries (described in detail in Chapter 6). Therein, due to the feature that the seasonal differences of the probabilistic characteristics of typhoon events are considered, it is possible to estimate portfolio losses in a certain period in a year. This is useful in practice when the assessments of portfolio losses are required for the remaining period of a year.

Another potential application of the developed model is the investigation of the effects of the global climate change on the probabilistic characteristics of strong wind speed induced by typhoons (described in detail in Chapter 7). A preliminary study on this is undertaken in (Graf et al., 2008). In this study, assuming several scenarios on the future change of SST, it is found that the upper quantile values of the annual maximum wind speed distribution may significantly increase as a consequence of an increase of the SST; in turn, the probability of failure of structures due to wind loads may also significantly increase. In a second study the typhoon model is re-established using the output of a climate model as a possible future dataset to investigate the change of the risk due to climate change.

Another application concerns real-time decision making (described in detail in Chapter 8) e.g. in the context of evacuation of people and shut-down of operation of engineered facilities in the face of an approaching typhoon, see (Nishijima et al., 2009). For this, the feature that the entire life of a typhoon is modelled is useful for simulating possible tracks and changes of the intensity of the approaching typhoon. Furthermore, the consideration of the SST and the seasonal differences of probabilistic characteristics of typhoon events enables utilizing additional information such as the current SST around the location of the typhoon and current season; consequently, the uncertainties associated with the transition of the typhoon can be reduced and decisions may be made more precisely.

2.3. Components of the typhoon model

The typhoon model developed during this thesis consists of two parts; a hazard model and a vulnerability model. The hazard model is composed of four components, i.e. occurrence model, transition model, wind field model and surface friction model, see Figure 2.1. The hazard model describes the probabilistic nature of the entire life-time of typhoons from their occurrence to dissipation. The occurrence model represents the probabilistic characteristics of the location and frequency of the initiation of typhoon events as a function of season and sea surface temperature (SST). The transition model concerns the probabilistic representation of the movement and change of intensities taking into account the spatial in-homogeneity of the probabilistic characteristics, seasonal differences and SST. The wind field model describes the wind fields induced by typhoons as a function of the relevant parameters characterizing the states of the typhoons. The surface friction model represents, as a function of surface roughness, the relation between the wind speed at the surface and the wind speed at gradient height at which the wind fields induced by typhoons are modeled. Finally, the vulnerability model represents the loss ratio of given types of exposure as a function of the wind speed acting on the exposures. In the next chapter, the datasets utilized for developing these models are introduced and, in the subsequent chapters, the methodology for the development of the typhoon model and the underlying assumptions are explained in detail.

2. Typhoon model

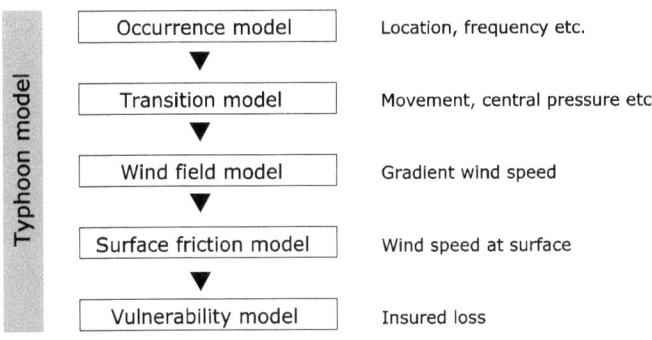

Figure 2.1: Components of the developed typhoon model.

To fulfill the requirements of the proposed Bayesian framework for probabilistic modelling of typhoon risks, a probabilistic typhoon model is established; the model is composed of the following components:

- A new occurrence model is developed, which represents the occurrence of a typhoon as a function of the location, the season (month) and the SST.
- For the transition model the proposed approach from (Vickery et al., 2000) is followed, but the model is adapted for the region of the North West Pacific and modified so that the development of a typhoon is also a function of the season (month) and the SST.
- The wind field model is established as proposed by (Georgiou et al., 1983).
- The surface friction model is developed as proposed by (Meng et al., 1997). A new scheme for the estimation of the roughness length is developed which combines the two approaches described in the state of the art (see Section 1.4).
- A model to estimate the portfolio losses is developed containing a new vulnerability model which considers the epistemic uncertainties (explained in Section 6.1).

2.4. Utilized datasets

The datasets utilized for developing the typhoon model are summarized in Table 2-1. The table contains the official titles of the datasets, the abbreviation of the title of each dataset that is used in this technical note, the components of the models established using each dataset, the period during which the observations in the datasets are recorded and the name of the provider of each dataset.

Table 2-1: Summary of utilized datasets.

Dataset title	Abbreviation	Model*	Period	Provider
Best track data of RSMC Tokyo-Typhoon Center	Best track data	O and T	1951-2006	JMA**
Oceanographic normals and analyses	SST data	O and T	1971-2000	JMA**
JMA annual observation report	Meteorological data	W and S	1961-2005	JMA**
Tochiriyou 3ji-mesh data***	Land-use data	S	2006	MLIT****
Digital Map 50m Grid (Elevation)	Elevation data	S	2008	JMC*****
Exposure data	Exposure data	V	******	Aon Benfield Japan
Loss data		V	******	Aon Benfield Japan

2. Typhoon model

Loss data
* O, T, W, S and V abbreviate occurrence model, transition model, wind field model, surface friction model and vulnerability model respectively. ** JMA abbreviates the Japan Meteorological Agency. *** No official title is available in English. **** MLIT abbreviates the Ministry of Land, Infrastructure, Transport and Tourism. ***** JMC abbreviates the Japan Map Center. ****** The period from which data is available differs between the data sets.

The best track data contains the information on the following typhoon characteristics at discrete times during the historical typhoon events:

- Storm category
- Location of typhoon (longitude and latitude)
- Central pressure
- Maximum 10-miniute sustained wind speed
- Direction of the longest radius of 50-knot wind
- The longest radius of 50-knot wind
- Direction of the shortest radius of 50-knot wind
- The shortest radius of 50-knot wind
- Direction of the longest radius of 30-knot wind
- The longest radius of 30-knot wind
- Direction of the shortest radius of 30-knot wind
- The shortest radius of 30-knot wind

Depending on the location of the typhoons and the occurrence year of the typhoon events the time interval of the records varies between 6 hours and 1 hour. The typhoon characteristics are recorded at shorter time intervals when typhoons are closer to the Japanese islands with higher intensities. The radiuses of 30-knot and 50-knot winds are available from 1977.

-41-

The SST data contains the information on the following parameters:

- 10-day mean SSTs in the northwestern Pacific ocean
- Monthly mean SSTs in the northwestern Pacific ocean
- Monthly mean SSTs in the oceans
- Monthly mean subsurface temperatures at the depth of 100 meter at seas around Japan
- Comparison between the SST means in 1971-2000 and the SST means in 1961-1990
- Coastal water temperature
- Long-term variations of SST

For the development of the occurrence model and the transition model, only the 10-day mean SST observations are utilized.

The meteorological data contains the information on the following atmospheric characteristics measured at meteorological stations (156 stations as of 2005):

- Temperature, vapor pressure and relative humidity
- Atmospheric pressure
- Wind speed and wind direction
- Cloud cover
- General weather condition
- Amount of evaporation
- Amount of global solar radiation
- Duration of sunshine
- Precipitation
- Snow cover and snow fall

2. Typhoon model

Whereas this dataset contains a variety of statistics of these parameters, e.g. daily maximum, monthly average etc. the statistics utilized in the development of the wind field model and the surface friction model are the following:

- 10-minute sustained wind speed
- Maximum wind speed
- Atmospheric pressure adjusted to sea surface level

These parameters are recorded on an hourly basis.

The land-use data contains the information on the use of land at each grid on the 100m-by-100m grid system defined by the MLIT. The land use is differentiated by the following categories:

- Rice field
- Plowed field
- Fruit farm
- Tree farm
- Woodland
- Waste land
- Building zone (sub urban)
- Building zone (city)
- Main line traffic zone
- Other land
- Lakes
- River zone (unused)
- River zone (artificial used)
- Beach
- Seawater
- Golf

2. Typhoon model

For each grid, the land-use category is recorded. The categories are utilized for identifying the roughness length, see Section 2.10.3.

The elevation data contains the information about elevation at each grid on the 50m-by-50m grid system defined by the JMC. The elevation data is used to estimate the topological factor, see Section 0.

The exposure data and loss data provided by Aon Benfield Japan are utilized to develop the vulnerability model. The data of each client is utilized only for the purpose of developing the vulnerability model and the data is not utilized for developing the other models. The content of the exposure and the loss data is explained in detail in Section 6.1.2. A confidentiality agreement between Aon Benfield Japan and ETH Zurich has been contracted. The confidentiality agreement includes that the data provided by Aon Benfield Japan and all results which are obtained by this data are made anonymous and censored in this thesis.

2.5. Occurrence model

The occurrence model describes the probabilistic characteristics of the initiation of typhoon events. Since the probabilistic characteristics vary with location and season it is required to consider the non-homogeneity of the characteristics in terms of time and space. The developed occurrence model represents the occurrence of a typhoon as a function of the location, of the season (month) and of the SST.

Within the present framework for risk management the entire lives of typhoons are modeled. This approach also enables a conditional simulation for the case of a approaching typhoon, since the information on the current state of typhoon can be used to simulate possible development of the typhoon.

2. Typhoon model

2.5.1. Definition of the initiation of typhoon events

Typhoons are an alias of intensified tropical storms which occur in the northwest Pacific region. Typically, tropical storms are not categorized as typhoons at their occurrences; only a few tropical storms develop and eventually become to be categorized as typhoons. The best track data provided by the JMA contains the records of all the tracks of the historical tropical storms that during their life-times are categorized as typhoons. It contains the records of the tracks of these tropical storms not only during the periods when the tropical storms are being categorized as typhoons but also during the periods before and after the storms are categorized as typhoons; however, the criteria are not clearly set for the condition under which the track of tropical storms begins to be recorded. In contrast, it is possible that the historical tropical storms that occurred but dissipated before they became to be categorized as typhoons may not be recorded in the best track data. Thus, defining the initiations of the typhoon events, in the development of the occurrence model, in accordance with the location at which the track of individual tropical storms starts to be recorded in the best track data may lead to a biased estimation of the probabilistic characteristics of the initiation of typhoon events. On the other hand, if the definition of the initiation of typhoon events was assumed given in accordance with the JMA definition – typhoons are defined as tropical storms which induce the wind speed equal or larger than 17.2 [m/s] – a number of track records would be disregarded, which are useful for establishing the transition model. Thus, in the development of the occurrence model the initiation of typhoon events is, by trial and error, assumed to be defined by the moment at which the central pressure of each tropical storm becomes less than 1000 [hPa] for the first time in the life-time of the tropical storm. If the track record of a tropical storm starts with the central pressure being less than 1000 [hPa] the first record of the tropical storm track is assumed to represent the initiation of the typhoon

2. Typhoon model

event[1]. The geographical distribution of the location of the initiation of typhoon events thus obtained is shown in Figure 2.2.

The initial condition of the historical typhoon events, i.e. location, translation direction and speed, and central pressure of typhoons at the time of the initiation of typhoon events and several times before the initiation are also obtained from the best track data, see Section 2.6 for the initial condition required for the simulation of typhoon events.

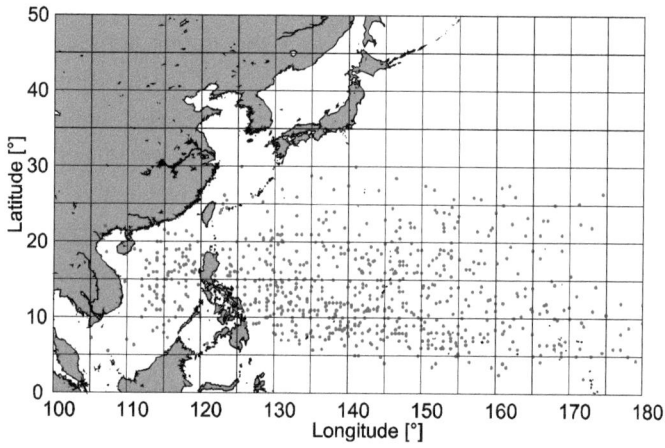

Figure 2.2: Geographical distribution of the location of the initiation of historical typhoon events.

2.5.2. Probabilistic model for the occurrence of typhoons

The development of the probabilistic model for the occurrence of typhoons considering SST changes is examined using Bayesian probabilistic networks, see Figure 2.3 as an example. A Bayesian probabilistic network consists of nodes and edged links; the nodes in the network represent (random) variables and the directed edges represent the probabilistic dependence of the variables. For instance, in the network shown in Figure

---

1 This is the case for 164 out of 1488 historical typhoon events considered in the development of the typhoon model.

2. Typhoon model

2.3 the probabilistic characteristics of the random variable *Occurrence* is defined as a function of the states of the *SST*, *Latitude* and *Longitude* and the probabilistic characteristics of the random variable *SST* is defined as a function of the states of the *Latitude*, *Longitude* and *Month*. The quantitative probabilistic dependencies of all the variables are estimated using EM-Learning algorithm (Hugin, 2006) and the using the historical data, i.e. the best track data and the SST data. The probability density and the probabilistic dependencies for each variable is represented by an empirical distribution.

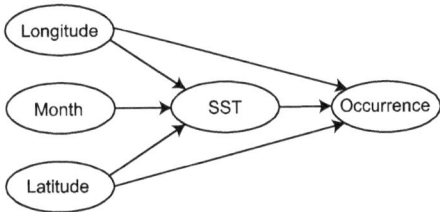

Figure 2.3: Bayesian probabilistic network for the occurrence of typhoons.

Presently, it is possible to simulate the occurrence of typhoons in consistency with historical data, see Section 3.1. However, further investigation is required for the cases where the SST is assumed to be increased to the extent that no or not sufficient historical observations are available. For this, references and comparisons to the outcomes from other scientific works are needed, which rely on several physics-based climate models, see the report published by the Intergovernmental Panel on Climate Change (Solomon et al., 2007) for overview. Furthermore, the probabilistic model for the initial states of typhoons given the occurrence of the typhoons must be developed.

Thus, the occurrence model that is developed is not employed; instead, historical observations of the initiation of the historical typhoon events, i.e. location of the occurrence and the initial state of the typhoons, are utilized (resampled) in the assessment of portfolio losses. Nevertheless, the comparison of the simulation results of the occurrence of typhoons using the occurrence model developed and the historical observations is shown in Section 3.1.

2.6. Transition model

The transition model describes the probabilistic characteristics of the movement of typhoons and the change of the intensity of typhoons. The variables considered in the transition model are; the translation speed and direction, central pressure and radius of maximum wind speed of typhoons. Therein, the non-homogeneity of the characteristics of these variables in terms of space and season are considered: The probabilistic model for the translation is developed for each month (from January to December) and for each 5°-by-5° grid over the entire northwest Pacific area (see Figure 2.4) and the probabilistic model for the central pressure is developed for each month and for each of the 18 different zones (see Figure 2.5) over the entire northwest Pacific area. Furthermore, the models for the translation are developed for the typhoons which are travelling eastwards and westwards respectively for the grids located at latitude equal to or lower than 30°N. These gridding and zoning are established by trial and errors. In contrast, the radius of maximum wind speed is modelled as a random variable, which is applied for all the months and the areas close to the Japanese islands[2].

---

[2] The assumption here is that the radius of the maximum wind speeds of a typhoon remains constant during the period when the typhoon is travelling close to the Japanese islands.

2. Typhoon model

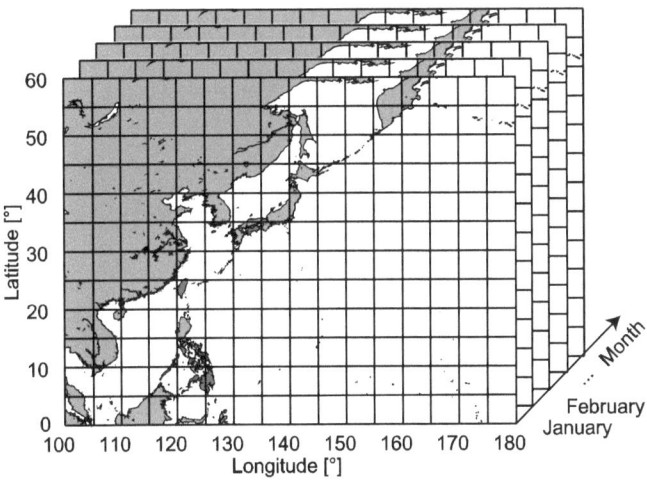

Figure 2.4: Grids and months for the probabilistic model for translation.

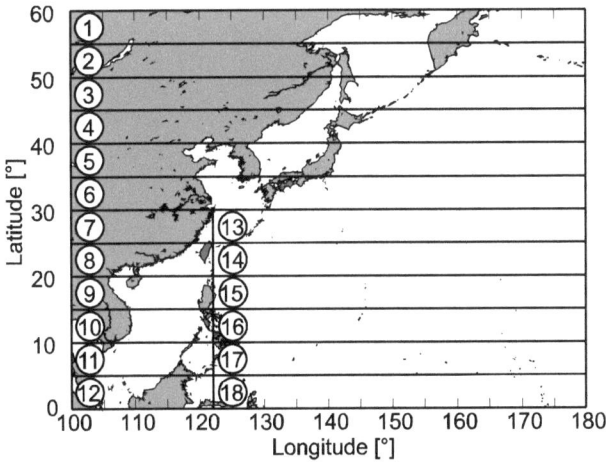

Figure 2.5: 18 zones for the probabilistic model for central pressure.
(The model is developed for each individual month in the same way as the model for translation, but it is not illustrated in the figure.)

2. Typhoon model

2.6.1. Probabilistic model for the translation of typhoons

The translation of typhoons is modeled by:

$$\Delta \ln V_i = a_1 + a_2 \ln V_i + a_3 \Gamma_i + \varepsilon_V \tag{2.1}$$

$$\Delta \Gamma_i = b_1 + b_2 V_i + b_3 \Gamma_i + b_4 \Gamma_{i-1} + \varepsilon_\Gamma \tag{2.2}$$

where V_i is the translation speed [km/h] at time step i, Γ_i is the translation direction [°] measured clockwise from north in the range of $(-180°, 180°]$ at time step i. $\Delta \ln V_i$ and $\Delta \Gamma_i$ are the difference of each respective quantity at the subsequent time steps, i.e. $\Delta \ln V_i = \ln V_{i+1} - \ln V_i$ and $\Delta \Gamma_i = \Gamma_{i+1} - \Gamma_i$. The time interval of the subsequent time steps is equal to 6 hours[3]. The coefficients $\mathbf{a} = (a_1, a_2, a_3)^T$, $\mathbf{b} = (b_1, b_2, b_3, b_4)^T$ are constants for a given grid and month. ε_V and ε_Γ are the residual terms, which are random variables representing the random fluctuations of the translation speed and angle. The residual terms ε_V and ε_Γ are assumed to follow the normal distribution with the mean value equal to zero and the standard deviations equal to σ_{ε_V} and $\sigma_{\varepsilon_\Gamma}$ respectively. Since the coefficients are estimated separately for eastwards ($\Gamma = (0, 180]$) and westwards ($\Gamma = (-180, 0]$) moving typhoons, it is possible that this model approach lead to a small discontinuity at 0° and ±180°. The model is continuous at 0° and an investigation of the model results shows that the discontinuity at ±180° is so small that it can be neglected, since the parameters are estimated and empirical calibrated using real data.

The coefficients $\mathbf{a}$ and $\mathbf{b}$ and the standard deviations σ_{ε_V} and $\sigma_{\varepsilon_\Gamma}$ are estimated by statistical analyses using the best track data for each individual grid and month and for each travelling direction, i.e. eastwards

---

[3] In the simulation of typhoon events, the transition of typhoons is first simulated with the time interval of 6 hours using the transition model. Then, the simulated states of the typhoons, i.e. location, translation speed and central pressure, are linearly interpolated with the time interval of 10 minutes; the wind field of the typhoon is simulated using these interpolated states with the time interval of 10 minutes.

2. Typhoon model

or westwards, if the grid is located at latitude equal to or lower than $30°$ N. In the case where sufficient data is not available for the statistical analyses for a particular grid and month, the values of the coefficients and the standard deviations estimated for one of the neighboring grids are substituted to the coefficients and the standard deviations for the grid and month. Thereby, the neighboring grid is chosen as such has the largest number of data in the neighboring grids for the month. Figure 2.6 shows the number of available data for the month July, August and September for eastwards and westwards moving typhoons.

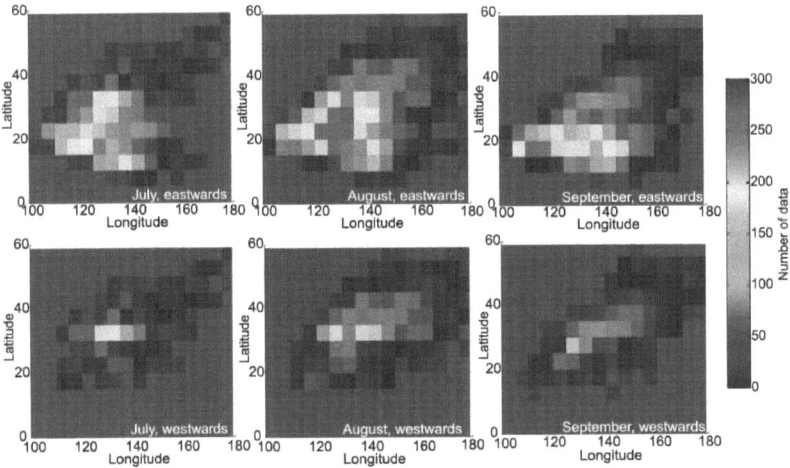

Figure 2.6: Number of data

Some criteria were introduced such that they prohibit the realizations of unrealistic translation speed in the simulation. The thresholds in the criteria were determined based on the historical data. Figure 2.7 shows the translation speed of the historical typhoons at different latitudes for the month September and the thresholds (represented by the red line) were set to 60 km/h for latitude below 30° and 120 km/h above 30°.

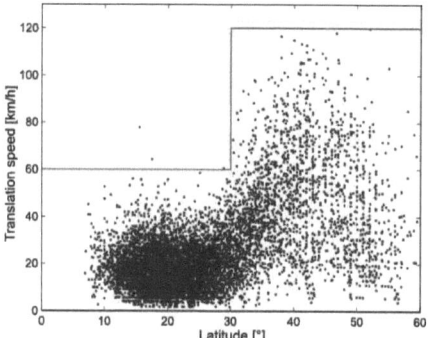

Figure 2.7: Translation speed

2.6.2. Probabilistic model for the central pressure of typhoons

The change of the central pressure when typhoons are at sea is modeled by:

$$P_{C,i+1} = c_1 + c_2 P_{C,i} + c_3 P_{C,i-1} + c_4 P_{C,i-2} + c_5 T_i + c_6 \Delta T_i + \varepsilon_{P_C} \qquad (2.3)$$

where $P_{C,i}$ is the central pressure at time step i, T_i is the SST at the location where the considered typhoon is located at time step i and $\Delta T_i = T_{i+1} - T_i$. The time interval of the subsequent time steps is the same as in the models for the translation and is equal to 6 hours. The coefficients $\mathbf{c} = (c_1, c_2, c_3, c_4, c_5, c_6)^T$ are constants for a given zone and month. ε_{P_C} is the residual term, which is a random variable representing the random fluctuation of the change of the central pressure.

The coefficients $\mathbf{c}$ are estimated by statistical analyses using the best track data[4] and SST data. The cumulative distribution function of the residual term ε_{P_C} is estimated by the empirical distribution; i.e. the realizations of the residual term, which are calculated using the estimated coefficients and the data, are used to represent the distribution of ε_{P_C}[5]. This is because, by visual inspections of the realizations of the residual terms, it seems that the

---

4 The records of the track of typhoons that are at sea are used for the statistical analyses.
5 In the simulation of typhoon events, these realizations of the residual terms are randomly re-sampled.

2. Typhoon model

distributions of ε_{p_c} significantly differ for different zones and months and any one of the common distribution families does not fit to the distribution of ε_{p_c} for all zones and months. The model for the central pressure is established for all the relevant zones and months without the substitutions as are employed in the case of the development of the model for the translation[6]. Note that since the SST data are not available for the periods of 1951-1970 and 2001-2006, the mean value of the SSTs at each location and month during the period of 1971-2000 is assumed to represent the SSTs for these periods and these SSTs are used for the statistical analyses.

2.6.3. Interpolation of the typhoon tracks

As described in Section 2.6.1, the typhoon track is estimated in 6h time steps. The discretization in 6h time steps has two disadvantages. First, it can happen that the simulated typhoon track points are not necessarily the points of the typhoon track which were the closest to the selected location and the track point which produces the highest wind speed for the selected location is missed. Second, it is possible that a simulated typhoon crosses the Japanese Islands within 6 hours. This would imply that the simulation produces a track point before and a track point after the Japanese islands, without applying the filling model in between. To overcome these drawbacks the typhoon tracks are linearly interpolated in 10min time steps if the typhoon is on land or close to the Japanese Islands as shown in Figure 2.8.

---

[6] In the case where the models are not established due to insufficient number of the data for the statistical analyses - which occurs only in irrelevant zones and months in the assessment of portfolios in Japan - the central pressure of typhoons is assumed not to change in these zones and months.

2. Typhoon model

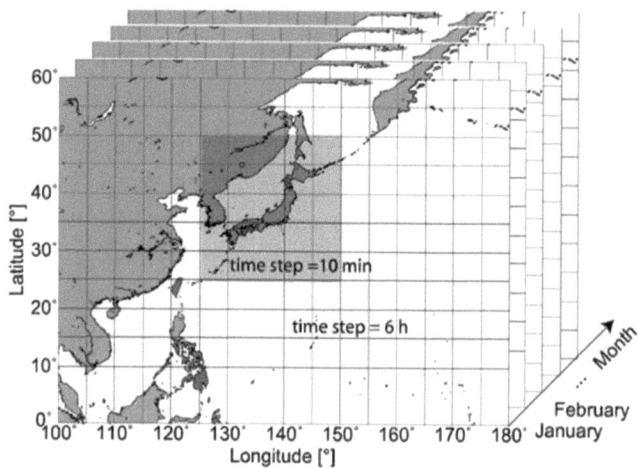

Figure 2.8: Interpolation area

2.6.4. Typhoon lysis

It is assumed, that a typhoon is dissolved when the central pressure is equal or higher than the peripheral standard pressure (here, 1013 [hPa] is assumed) or if the typhoon leaves the considered area as shown in Figure 2.4.

2.7. Filling model

The model thus established for the central pressure can be applied only when typhoons are at sea; after a typhoon makes landfall on the Japanese islands, the filling model explained below is applied until the typhoon either dissipates or passes through the land and is at sea again.

The filling model is modeled as:

$$\Delta P_t = \Delta P_0 \cdot \exp\left(-(d_1 + d_2 \Delta P_0)t\right) \tag{2.4}$$

where ΔP_0 is the difference of the central pressure of a typhoon at the moment of the landfall and the peripheral pressure (here, 1013 [hPa] is assumed), ΔP_t is the difference of the central pressure of the typhoon at time t and the peripheral pressure, and t is the time [hour] elapsed since the landfall. The coefficients $\mathbf{d} = (d_1, d_2)$ are constants and are estimated by statistical analyses using the best track data[7].

The uncertainties associated with the assumed functional form and the estimations of the coefficients are not considered in the simulation of typhoon events; that is, the deterministic function (2.4) with the estimated coefficients is utilized in the simulation of the typhoon events. Further investigation is required for the modelling of the change of the intensity of typhoons when they are on lands. This is addressed as one of future tasks.

---

[7] The records of the track of the typhoons which are on the Japanese lands are used for the statistical analyses.

2.8. Probabilistic model for the radius of maximum wind speed

The radius R_M of the maximum wind speed is assumed to be random but constant for each typhoon during the period when the typhoon is traveling around the Japanese islands. In order to estimate the probability distribution of R_M, first the radiuses of maximum wind speed of the historical typhoons are estimated.

The approach to estimate the radiuses of maximum wind speed of the historical typhoons is to employ the pressure field model proposed by (Schloemer, 1954a), which is utilized as part of the wind field model described in the subsequent section.

$$P_r = P_C + \Delta P \cdot \exp\left(\frac{r_M}{r}\right) \qquad (2.5)$$

Whereby r_M is the radius of maximum wind speed, r is the distance from the center of the typhoon to the considered location, P_C is the central pressure of the typhoon, P_r is the pressure at the considered location and $\Delta P = 1013 - P_C$ is the difference between the peripheral pressure and the central pressure.

Using both the observations for the pressure at the meteorological stations P_r and the central pressures of the typhoons P_C, the radius of the maximum wind speed r_M can be estimated by linearizing the equation and using a linear regression:

$$\ln\left(\frac{P_r - P_C}{\Delta P}\right) = -\frac{1}{r} \cdot r_M \qquad (2.6)$$

However, this approach can fail when typhoons are relatively far away from the Japanese islands since the number of the observations for the pressure at the meteorological stations that can be used for the statistical analysis is limited; for this reason only the data of the typhoons which were close to the Japanese Islands (in a radius of 250km) were used to estimate the probability distribution of R_M.

Using this approach the radius of the maximum wind speeds is estimated for all typhoons after 1970 which made landfall to the Japanese Islands. Based on these estimated values the radius of maximum wind speed is represented with a truncated lognormal distribution with the parameters $\lambda = 4.8$, $\varsigma = 0.4815$ and the distribution is truncated below 30 km and above 400 km.

2.9. Wind field model

The wind field model developed in this thesis is a deterministic model which describes the wind fields as a function of the relevant variables of the state of a typhoon. The variables required to describe the wind field are the central pressure p_C, the radius r_M of maximum wind speed, the translation speed v and translation direction γ of the typhoon.

2.9.1. Model for pressure fields

Using the central pressure p_C and the radius r_M of maximum wind speed of a typhoon at a given time, the pressure field is modeled as proposed by (Schloemer, 1954a) as:

$$p(r) = p_C + \Delta p \cdot \exp\left(-\frac{r_M}{r}\right) \tag{2.7}$$

where $p(r)$ is the pressure at the location whose distance from the center of the typhoon is r and Δp is the difference between the central pressure and the peripheral pressure, i.e. $\Delta p = 1013 - p_C$ [hPa].

2. Typhoon model

2.9.2. Model for wind field as a function of pressure field

The wind field induced by the typhoon at the time is modeled as proposed by Georgiou et al. (1983) and applied in other models, see e.g. (Meng et al., 1995a) using the pressure field $p(r)$ represented by Equation (2.7) as:

$$\tilde{u}_g(r,\alpha) = \frac{v\sin\alpha - fr}{2} + \sqrt{\left(\frac{v\sin\alpha - fr}{2}\right)^2 + \frac{r}{\rho}\frac{\partial p(r)}{\partial r}} \qquad (2.8)$$

where $\tilde{u}_g(r,\alpha)$ is the wind speed at gradient height at the location whose distance from the center of the typhoon is r and whose angle measured clock-wise relative to the translation direction of the typhoon is α, see Figure 2.9. $\partial p(r)/\partial r$ is calculated using Equation (2.7). f is the Coriolis parameter and ρ is the air density. Measuring the parameters in Equation (2.8) in terms of [kg] for mass, [m] for length and [s] for time, the Coriolis parameter is written as $f = 1.46 \times 10^{-4} \times \sin\phi$ [1/s] where ϕ is the latitude of the representative location[8] of the typhoon. $\rho = 1.275$ [kg/m^3] is adopted as the air density. Figure 2.10 shows an example of the wind field calculated using the wind field model, i.e. Equations (2.7) and (2.8).

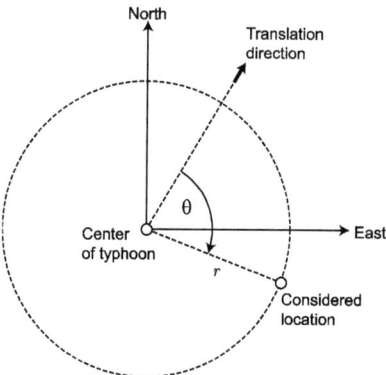

Figure 2.9: Coordinate system used in the wind field model.

---

[8] The centers of typhoons are used as the representative locations of the typhoons.

2. Typhoon model

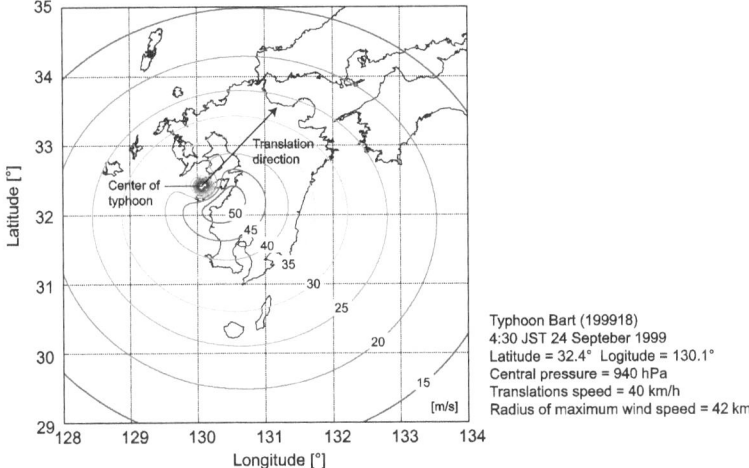

Figure 2.10: Example of the wind field calculated using the wind field model.

The wind speed $\tilde{u}_g$ calculated in accordance with Equation (2.8) corresponds to the instant wind speed at the moment of the time step of the typhoon simulation and does not correspond to the 10-minute sustained wind speed[9]. The wind speed will vary with time due to the change of the characteristics of a typhoon (e.g. position, central pressure, translation speed and direction). Therefore it is assumed that the wind speed $\tilde{u}_g$ corresponds to a wind speed averaged during a one-hour period or longer, see e.g. (Vickery and Twisdale, 1995c) and (Matsui et al., 1998). However, the exact time period during which the wind speeds calculated by the wind field model are considered to be averaged is not necessarily required for the development of the hazard model. What is required in the development of the hazard model is the relation between the wind speed calculated by the

---

9 In the present typhoon model the maximum of 10-minute sustained wind speeds during each typhoon event is used as the hazard index. However, in principle other hazard indices may also be suitable. One of the reasons why the maximum of 10-minute sustained wind speeds is chosen as the hazard index in the present typhoon model is that it allows for the direct comparison of the wind hazard map established using the present typhoon model with the wind hazard map provided by the Architectural Institute of Japan Architectural Institute of Japan (2004). *AIJ Recommendations for Loads on Buildings..*

2. Typhoon model

wind field model and the 10-minute sustained wind speed at surface (as a function of land-use as explained in section 2.10); this relation can be empirically established by comparing the wind speeds reproduced using the wind field model and the observed 10-minute sustained wind speeds at meteorological stations during the historical typhoon events.

2.9.3. Calculation of maximum 10-minute sustained wind speeds

The meteorological data contains the records of the 10-minute sustained wind speeds only for the first 10 minutes every hour. Thus, it is possible that the *true* maximum 10-minute sustained wind speed during a historical typhoon event may be larger than the maximum of the 10-minute sustained wind speeds during the event, which are recorded in the meteorological data. In order to take this into consideration in the assessment of the maximum 10-minutes sustained wind speeds using the hazard model, the following logic is assumed.

First, it is assumed that each 10-minute sustained wind speed recorded in the meteorological data represents the wind speed over the next one hour; i.e. the recorded 10-minute sustained wind speeds are assumed to be equal to the one-hour sustained wind speeds. Hence, the surface friction model described in Section 2.10 represents the relation between the wind speeds calculated by the wind field model and the *one-hour sustained* wind speeds at surface.

Then, the one-hour sustained wind speeds are related to the maximum of the 10-minute sustained wind speeds during the corresponding one-hour periods as follows. Assuming that six non-overlapping 10-minute sustained wind speeds in one hour are independent and identically distributed and each of them follows a normal distribution with mean value equal to $\tilde{u}_{s,60}$ and standard deviation equal to $\sigma_{u_{s,10}}$, the cumulative distribution function of the maximum $U_{s,10}$ of the six 10-minute sustained wind speeds is obtained as:

$$F_{U_{s,10}}(u) = \left\{ \Phi\left(\frac{u - \tilde{u}_{s,60}}{\sigma_{u_{s,10}}}\right) \right\}^6 \tag{2.9}$$

where $\Phi(\cdot)$ is the standard normal cumulative distribution function, $\tilde{u}_{s,60}$ is the one-hour sustained wind speed and $\sigma_{u_{s,10}}$ is the standard deviation of the 10-minute sustained wind speed at surface. By substituting into Equation (2.9) $\sigma_{u_{s,10}} = 2.6$ [m/s], which is estimated for strong winds at surface using the detailed records of the wind speeds during several historical typhoons by (Matsui et al., 1998), the mean value of the maximum of the six 10-minute sustained wind speeds is approximated as:

$$E[U_{s,10}] \approx \tilde{u}_{s,60} + 3.3 \text{ [m/s]} \tag{2.10}$$

Finally, in the simulation of typhoon events, (1) the wind speeds at gradient height calculated by the wind field model are converted to the one-hour sustained winds using the surface friction model (see Section 2.10) and then (2) the one-hour sustained wind speed at surface are converted to the mean 10-minute sustained wind speeds at surface using Equation (2.10), see Figure 2.11. These conversions of the wind speeds are made for each location. In the following, the mean value $E[U_{s,10}]$ is referred to as the 10-minute sustained wind speed at surface calculated by the hazard model unless stated otherwise.

2. Typhoon model

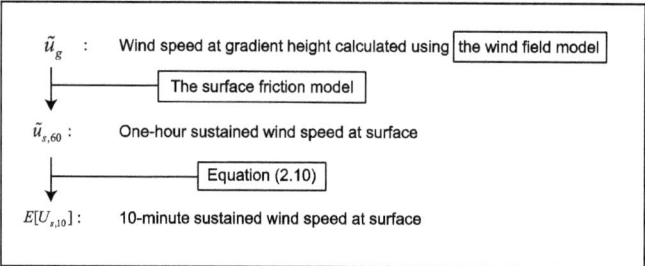

Figure 2.11: Conversions of wind speeds ($\tilde{u}_g$ is the abbreviation of $\tilde{u}_g(r,\alpha)$ for any given location).

2.9.4. Wind direction

The wind direction is taken into account by the following assumption. The wind direction at gradient height at a considered location is the tangential vector (counter clock wise) at the circle with the center of the typhoon as center and the distance between the center of the typhoon and the considered location as radius of the circle as shown in figure 2.10.

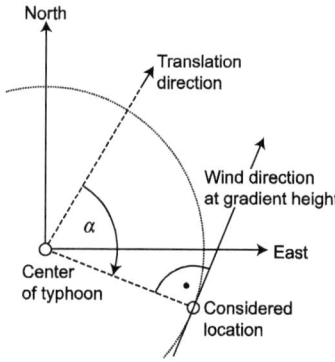

Figure 2.12: Coordinate system used in the wind field model.

2. Typhoon model

The consideration of wind direction facilitate to assess wind speeds at surface in more detail since different profiles of surface in different directions can be taken into account for the conversion of wind speeds at gradient height to wind speeds at surface (See Section 2.10). The process for estimating the wind direction at surface level is explained in Section 2.10.2.

2.10. Surface friction model

The surface friction model developed describes the relation between the wind speeds at gradient height and the wind speeds at nominal height on surface and the relation between the wind directions at gradient height and the wind directions at nominal height on surface. The definition of the nominal height is explained later. Further, the surface friction model developed in this thesis is deterministic.

2.10.1. Relation between the wind speeds at gradient height and at nominal height

Taking basis in Davenport (1965) and Meng et al. (1997) the relation between the wind speed at gradient height and the wind speed at nominal height is assumed to be represented as:

$$u(z) = \tilde{u}_g \left(\frac{z}{z_g}\right)^a \cdot E_g \quad (2.11)$$

where $\tilde{u}_g$ is the wind speed at gradient height at any given location calculated by the wind field model (see Equation (2.8); here, the augment (α, r) is abbreviated), E_g is the topological factor (see the explanation in Section 0) z_g is the gradient height and $u(z)$ is the *one-hour* sustained wind speed at the height of z (see the explanation in Section 2.9.3). These heights are measured from the adjusted surface level defined by the following equation (Simiu et al., 1976), see Figure 2.13:

$$d = 0.75h \quad (2.12)$$

where h represents the average height [m] of roughness elements, e.g. buildings, at the considered area, which is written as; $h = Az_0^{0.86}$ (Lettau, 1970) and, $A = 11.4$ (Helliwell, 1971; Kondo and Yamazawa, 1986), see (Meng et al., 1995a). z_0 is the roughness length [m]. The nominal height is defined as the height of 10 [m] from the adjusted surface level. The exponent α and the gradient height z_g are assumed to be represented respectively as:

$$a = 0.27 + 0.09\log_{10} z_0 + 0.018(\log_{10} z_0)^2 + 0.0016(\log_{10} z_0)^3 \quad (2.13)$$

$$z_g = 0.052 \frac{\tilde{u}_g}{f_\lambda}(\log_{10} Ro_\lambda)^{-1.45} \quad (2.14)$$

where the modified Rossby number is written as $Ro_\lambda = \tilde{u}_g / (f_\lambda \cdot z_0)$, f_λ and ξ are the two parameters proposed by (Meng et al., 1997) for describing the structure of strong wind in the typhoon boundary layer and are given as:

$$f_\lambda = \left(\frac{\partial \tilde{u}_g(r,\alpha)}{\partial r} + \frac{\tilde{u}_g(r,\alpha)}{r} + f \right)^{\frac{1}{2}} \cdot \left(2\frac{\tilde{u}_g(r,\alpha)}{r} + f \right)^{\frac{1}{2}} \quad (2.15)$$

$$\xi = \left(2\frac{\tilde{u}_g(r,\alpha)}{r} + f \right)^{\frac{1}{2}} / \left(\frac{\partial \tilde{u}_g(r,\alpha)}{\partial r} + \frac{\tilde{u}_g(r,\alpha)}{r} + f \right)^{\frac{1}{2}} \quad (2.16)$$

2. Typhoon model

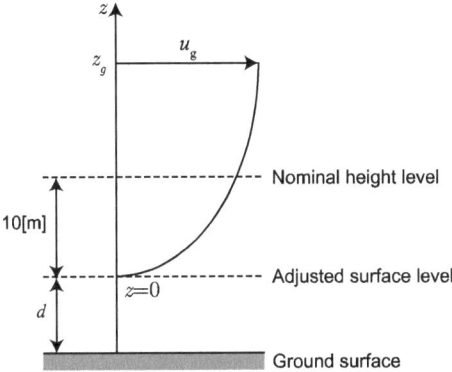

Figure 2.13: Coordinate system in vertical direction.

Thus, the exponent α and the gradient height z_g can be calculated given the value of the roughness length z_0 using Equations (2.12) - (2.16).

The procedure to identify the roughness length z_0 at each location on the Japanese islands is described in Section 2.10.3.

2.10.2. Relation between the wind direction at gradient height and at nominal height

Taking basis in Meng et al. (1997) the inflow angle, which is describing the difference between the wind direction at gradient height and the wind direction at nominal height, is assumed to be represented as:

$$\gamma(z) = \gamma_s \left(1 - 0.4 \frac{z}{z_g}\right)^{1.1} \qquad (2.17)$$

where $\gamma(z)$ is the inflow angle at the height z, z_g is the gradient height and with

$$\gamma_s = (69 + 100\xi)(\log_{10} Ro_\lambda)^{-1.13} \qquad (2.18)$$

where Ro_λ is the modified Rossby number and, f_λ and ξ are the two parameters proposed by (Meng et al., 1997) for describing the structure of strong wind in the typhoon boundary layer (see Equation (2.15) and (2.16)).

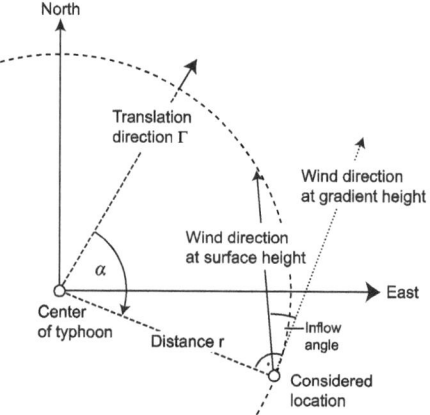

Figure 2.14: Coordinate system used in the surface friction model.

Figure 2.15 shows the wind field induced by the typhoon Songda (2004 18) at gradient height (left) and the wind field converted to surface height considering the roughness and the effect of the topography (right).

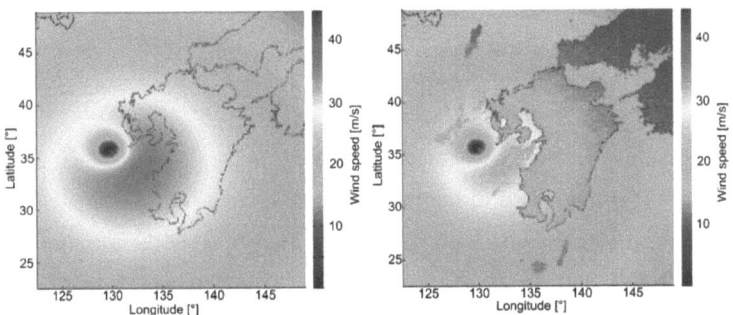

Figure 2.15: Wind field at gradient height (left) and wind field at surface height (right) of the typhoon Songda 2004 18.

2.10.3. Roughness length

The value of the roughness length z_0 is estimated for the surrounding area of the JMA stations using the gust factor (Grimmond and Oke, 1999; Verkaik, 2000) by applying the approach proposed by (Wieringa, 1986; Wieringa, 1993). In a next step the roughness lengths for the different land use categories are optimized by maximizing the correlation between the roughness length estimated from the gust factor and the roughness length estimated using the land use data. To estimate the roughness length for any selected location on the Japanese Islands, the following approach has been applied:

It is assumed that the vertical profile of the wind speed is influenced by the state of the roughness within the range of 3 km of the upwind terrain

(Wieringa, 1986). Following this assumption, the area around meteorological stations within a radius of 3 km is divided into 16 sectors with equal angles, and then the roughness length z_0 is estimated using the gust factor calculated out of the wind speed measurements as proposed by (Wieringa, 1986):

$$z_0 = z_s / \exp\left(-\frac{\left(1.42 \cdot 0.3\ln\left(10^3 / Ut - 4\right)\right)}{G-1}\right) \quad (2.19)$$

Where z_s is the height of the measurement device [m], U is the average wind speed (10min sustained wind speed [m/s]), t is the duration of the record [sec.] and G is the gust factor calculated as (Ashcroft, 1994; Vickery and Skerlj, 2005):

$$G = \frac{u_{gust}}{u_{10}} \quad (2.20)$$

Where u_{gust} is the gust wind speed and u_{10} is the 10-min sustained wind speed in m/s. Thereby, in order to exclude possible effects of local terrain on the wind speeds, such as steep slope, only the meteorological stations and the sectors where the tangent of the average slope is smaller than 7.5° are considered.

In order to estimate the roughness length not only for the surrounding area of the meteorological stations but for the entire Japanese Islands the GIS-based land-use data is utilized. The land-use data contains the information on the use of the land at each grid on the 100m-by-100m grid system defined by the MLIT. The land use is differentiated by the categories in Table 2-2.

In the land-use data from 2006 (compared to the previous years), there is no differentiation between the land-use category 7 "Building zone (sub urban)" and 8 "Building zone (city)" (see Figure 2.16). In order to separate these two categories the map of city centres (available at: http://nlftp.mlit.go.jp/) was used to determine which "Building zone" belongs to the city and which to the sub urban area (see Figure 2.17).

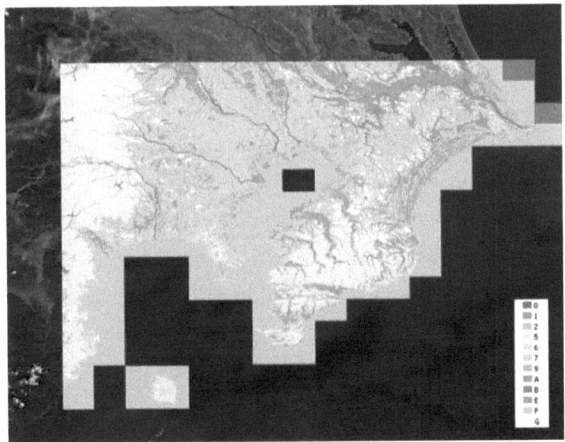

Figure 2.16: Land use data

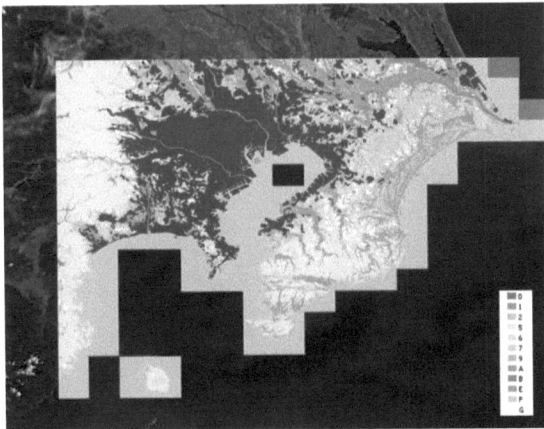

Figure 2.17: Land use data and map of city centers

Then, the "optimal" values of the roughness lengths for each of the land use categories are identified by maximizing the correlation coefficient between the roughness length estimated using the gust factor and the roughness length estimated using the land use data. The roughness length for a considered location using the land use data is estimated by the following procedure:

For estimating the roughness length for a considered location, the land use data within the circle segment with an angle of 22.5° and a radius of 3km of the upwind terrain is considered (Bartheelmie et al., 1993). This circle segment is divided into three areas as shown in Figure 2.18, in each of these three areas the average of the roughness length of the individual categories of the land use data is calculated. The average of the roughness length of these three areas is used as the roughness length for the considered location.

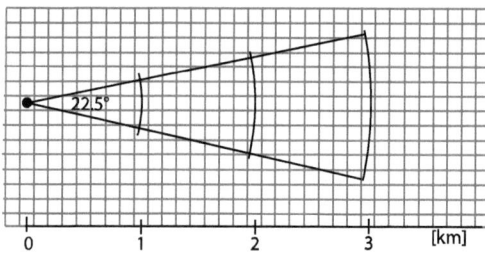

Figure 2.18:Considered segment for estimating the roughness length.

Table 2-2: Categories of the land use data and the according roughness length

Land use	Categories	Roughness length
Rice field	1	0.01
Plowed field	2	0.15
Fruit farm	3	N/A
Tree farm	4	N/A
Woodland	5	0.60
Waste land	6	0.01
Building zone (sub urban)	7	0.60
Building zone (city)	8	0.95
Main line traffic zone	9	0.30
Other land	A	0.10
Lakes	B	0.0001
River zone (unused)	C	N/A
River zone (artificial used)	D	N/A
Beach	E	0.12
Seawater	F	0.0001
Golf	G	0.13
No Data	H	0.0001

2.10.4. Topography

Topographic features, such as hills, escarpments and ridges, have strong effects on the wind speed profiles. These topographic features act as obstacles to the boundary layer and accelerating the wind near the surface. In many codes, there are methods to estimate this wind speed-up effect, a overview can be found at (Ngo and Letchford, 2008).

The effect of the wind speed-up due to the topography is represented with the topography factor E_g as proposed in the AIJ load recommendations (Architectural Institute of Japan, 2004) and was investigated in detail in (De Sanctis et al., 2008).

$$u(z) = \tilde{u}_g \left(\frac{z}{z_g}\right)^a \cdot E_g \tag{2.21}$$

$$E_g = (C_1 - 1)\left(C_2\left(\frac{Z}{H_s} - C_3\right) + 1\right)\exp\left(-C_2\left(\frac{Z}{H_s} - C_3\right)\right) + 1 \tag{2.22}$$

$$\theta_S = \tan^{-1}\frac{H_S}{2L_S} \tag{2.23}$$

Where C_1, C_2, C_3 are the parameters determining the topography factor and depending on the topography shape (see Appendix B 10.2), inclination θ_S and distance $X_s(m)$ from the top of the topographic feature to the construction site. When the inclination θ_S is greater than $60°$, the topography factor is assumed to be $60°$. $Z(m)$ is the height above ground. It is assumed to have the same value as Z_b when it is smaller than Z_b. $H_S(m)$ is the height of the topography and $L_S(m)$ is the horizontal distance from the top of topographic feature to the point where the height is half the topography height as shown in Figure 2.19 and Figure 2.20.

2. Typhoon model

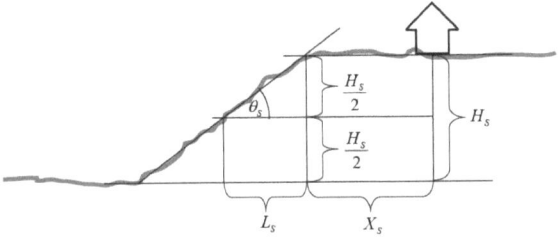

Figure 2.19: Escarpments

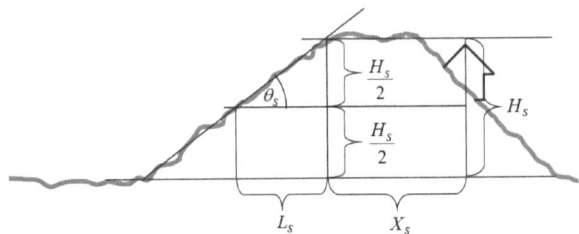

Figure 2.20: Ridge-shaped topography

For a particular inclination θ_S and a horizontal location X_S/H_S, the topography factor is calculated by linearly interpolating from the values at the nearest inclinations and horizontal locations.

The inclination θ_S and the horizontal location is obtained from the elevation data, which contains the information on elevation at each grid on the 50m-by-50m grid system defined by the JMC. Figure 2.21 shows a section of the elevation data in the area of Tokyo.

2. Typhoon model

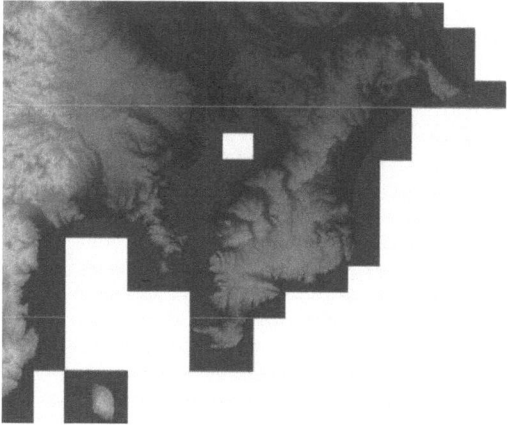

Figure 2.21: Topographic map

The topography factor E_g is calculated for each point on a 100m by 100m grid over entire Japan. For estimating the topography factor for a cell in the 1km by 1km grid the average of the 100 points within this grid is taken.

2.11. Software tool for creating a hazard event set using the hazard model

A hazard event set builder software tool is developed to create a stochastic event set using the typhoon model described in this chapter. This hazard event set builder is written in MATLAB and simulates (using a Monte Carlo simulation) typhoon events which correspond to a specified number of years. The output of this hazard event set builder are the simulated typhoon tracks and the corresponding wind speeds in the defined 1km by 1km grid over the Japanese Islands. A stochastic event set, which contains the typhoon events of 24'000 years, is established and used for the portfolio risk analysis described in Chapter 6. The hazard event set builder is also used to perform the simulations used for the studies described in Chapters 7 and 8.

3. Verification and validation of the typhoon model

All the individual components of the developed typhoon model and the typhoon model as a whole are verified. The verification here refers to checking the consistency of the outcomes from these component models with respect to the historical observations.

The validation of the developed typhoon model is done by establishing the typhoon model using only a part of the available historical observations. The statistics obtained from this established model is compared against the historical observations which aren't used for establishing the model.

3.1. Occurrence of typhoons

The simulation results of the occurrence of typhoons using the probabilistic model presented in Section 2.5 are compared with the historical observations. Figure 3.1 shows the occurrence rate of typhoons in each 1°-by-1° grid calculated using the historical data (left) and simulated using the probabilistic model (right). Whereas these figures show good agreement it is not verified to which extent the probabilistic model can be extrapolated to higher SST.

3. Verification and validation of the typhoon model

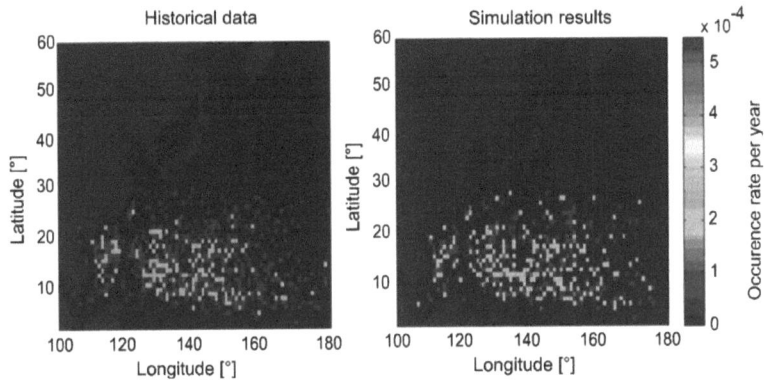

Figure 3.1: Comparison of the occurrence of typhoons.

3.2. Transition of typhoons

The simulation results of the transitions of typhoons are compared with the historical observations with respect to: (1) the frequency of the typhoons with certain intensities which cross certain latitudes, (2) translation angle and (3) translation speed of the typhoons at the moment when the typhoons cross these latitudes. Furthermore, the mean values of the number of the landfalls to the Japanese islands in each month are compared.

First, the cumulative frequencies of typhoons crossing different latitudes ($25°, 30°, 35°$ and $40°$) between longitudes $[120°, 160°]$ (see Figure 3.2) are compared as a function of the central pressure as shown in Figure 3.3. In the figure the comparison is shown for August and September. In the horizontal axis in the figure, a bar at "<950", for example, represents the mean frequency of the typhoons crossing the latitude with the central pressure smaller than 950 [hPa] at the moment when they cross the latitude. As it can be seen in the figure, the simulation results well represent the historical observations. The comparisons are made for other relevant months and they also show good agreements.

3. Verification and validation of the typhoon model

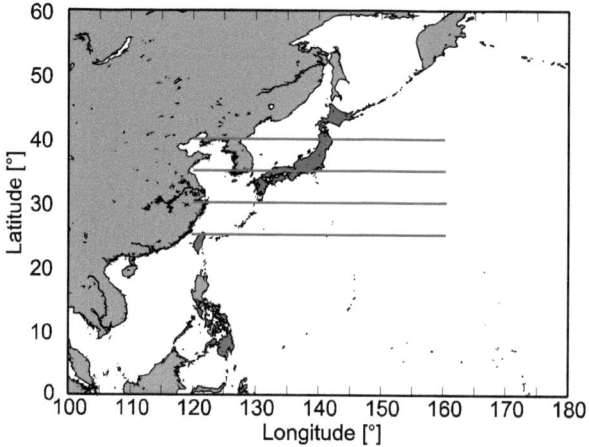

Figure 3.2: Lines and area which the probabilistic characteristics of typhoons travelling through are compared.

3. Verification and validation of the typhoon model

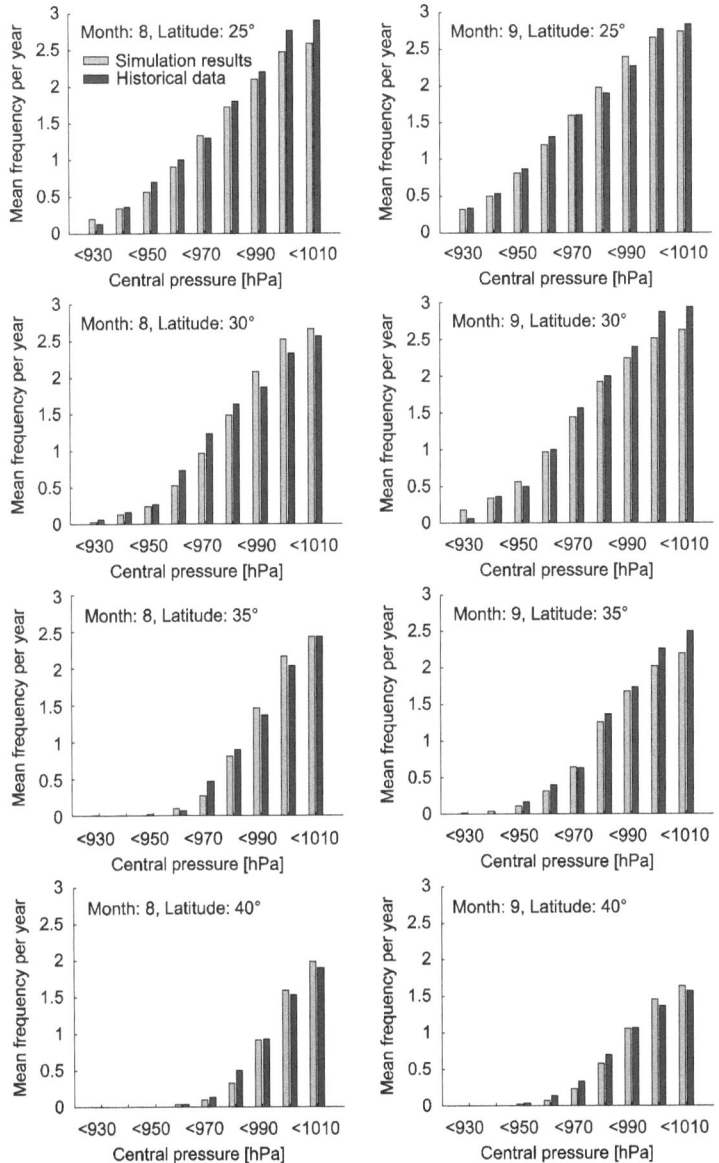

Figure 3.3: Cumulative mean frequencies of typhoons crossing different

3. Verification and validation of the typhoon model

latitudes between longitudes [120, 160°] for August and September.

Second, the probability distribution of the translation speed and direction of the typhoons at the moment when they cross certain latitudes is compared. Figure 3.4 shows the comparison in September and at latitude equal to 30°. As it can be seen in the figure, the simulation results well represent the historical observations. The comparisons are made for other relevant months and for the other latitudes and they also show good agreements. A comparison for different latitudes and different months is shown in the Appendix A 10.1. The variation increases at locations of higher latitudes. This can be explained by the fact that the difference in the state of typhoons is accumulated by the repetitive use of the different Markov models in the simulation.

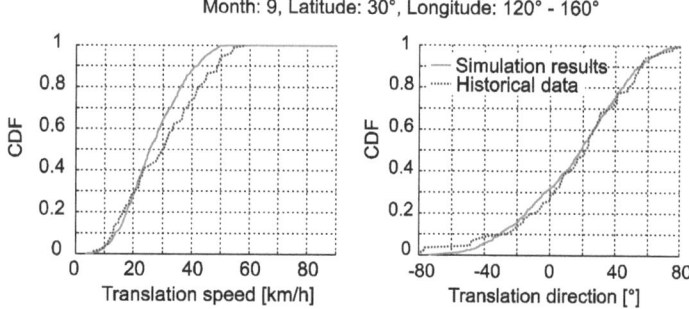

Figure 3.4: Cumulative distributions of the translation speed and direction of typhoons crossing the latitude of 30° between the longitudes [120, 160°] in September.

Finally, the mean frequency of the landfalls of typhoons to the Japanese islands in each month is compared. As it can be seen in the figure, the simulation results well represent the historical observations for relevant months. Note that the numbers of landfalls of typhoons in the simulation results are significantly higher in March to May, October and November; however, these typhoons are weak and thus do not significantly contribute to portfolio losses.

3. Verification and validation of the typhoon model

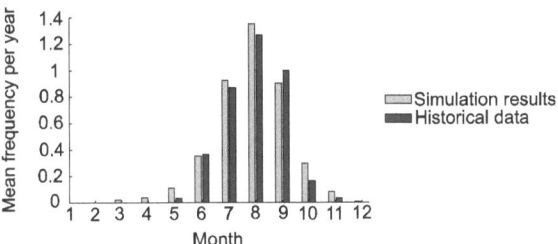

Figure 3.5: Mean frequency of the landfalls of typhoons in different months.

For the purpose of verifying and validating the developed occurrence and transition models, the mean annual numbers of the typhoons simulated using these models that intersect certain latitudes and make landfall to the Japanese islands are compared with those of the historical typhoons in two ways: Firstly, the occurrence and transition models are developed using historical data during the period from 1956 to 2006 and comparisons are made with the historical typhoons in the same period (the results are shown in Figure 3.6). Secondly, the occurrence and transition models are developed using the historical data of every-two years (odd years) during the period from 1956 and 2006 and comparisons are made with historical data of the other every-two years (even years) during the same period (the results are shown in Figure 3.7).

Figure 3.6 (left and centre) shows the comparisons of the statistics in the case of the month of September. In the figure, the mean annual numbers of the typhoons that intersect the latitudes of 30° and 35° within the range of the latitudes of 120° and 160° are compared as a function of the central pressure of the typhoons at the moment of the intersection. Figure 3.6 (right) shows the comparison of the mean annual numbers of the typhoons that make landfall to the Japanese islands in each month. These comparisons show good agreement, implying that the procedure for

3. Verification and validation of the typhoon model

estimating the parameters of the occurrence and transition models is appropriate; i.e. the developed occurrence and transition models are verified. Figure 3.7 shows the comparisons of the same statistics as are shown in Figure 3.6, which, however, are obtained based on the occurrence and transition models developed using the "odd years" data and based on the historical "even years" data. These comparisons also show good agreement, implying that the overall approach described in the present chapter for the development of the occurrence and transition models is not sensitive to the data utilized; the developed models are validated in this sense.

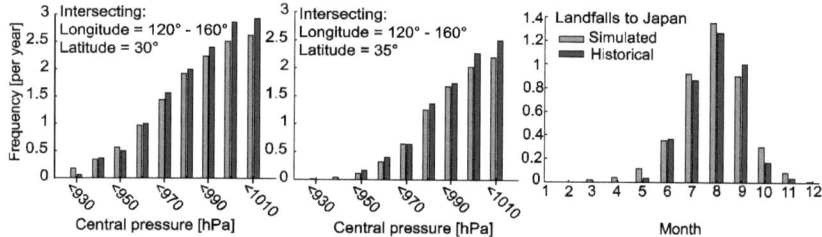

Figure 3.6: Verification of the procedure for the development of the occurrence and transition models.

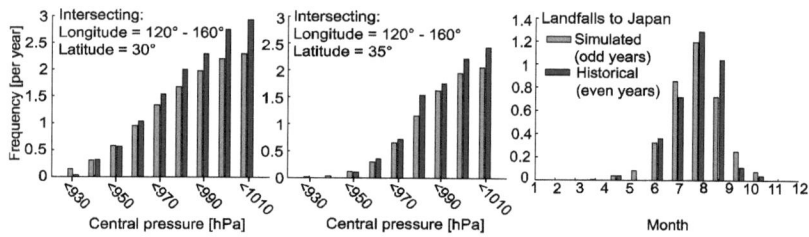

Figure 3.7: Validation of the approach employed for the development of the occurrence and transition models.

3. Verification and validation of the typhoon model

3.2.1. Extrapolation of the transition model to the future

In order to investigate if the developed typhoon model can be used to represent the characteristics of typhoons in the future, the occurrence and transition models are developed using the historical data of 1951 to 1978 and comparisons are made with historical data of the years 1979 to 2006 (the results are shown in Figure 3.8).

Figure 3.8 shows the comparisons of the same statistics shown in Figure 3.6, which, however, are obtained based on the occurrence and transition models developed using the first half of the available years of the data (1951 to 1978). These comparisons show acceptable agreement, although Figure 3.8 shows that the simulation results based on the models developed on the years 1951 to 1978 underestimate the number of typhoons with low central pressure. This implies that further investigations have to be done to analyze if there is a trend in the historical data, which has to be considered or if the insufficient number of data to establish the models (only 50% of the available data is used) is the reason for this discrepancy, before the typhoon model is used to predict future scenarios.

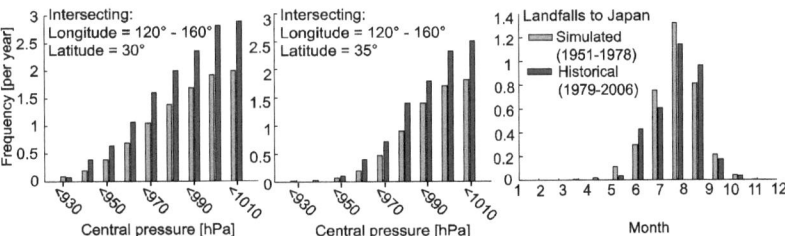

Figure 3.8: Extrapolation of the typhoon model to the future.

3. Verification and validation of the typhoon model

3.2.2. Seasonal differences in the transition model

To analyze the effect of the consideration of the seasonal difference, the occurrence and the transition model are developed considering and without considering the seasonal difference. The simulation results of the model considering the seasonal difference, the simulation results of the model without considering the seasonal difference and the historical data are compared in Figures 3.9 to 3.12.

The probability distribution of the translation speed and direction of the typhoons at the moment when they cross certain latitudes is compared. Figure 3.9 shows the comparison for all months and at latitude equal to 25° and 30°. As can be seen in the Figure 3.9, the simulation results of both models do not have a bias and both well represent the historical observations over all months.

The difference of the simulation results of the two models (one considering and one not considering the seasonal difference) can be seen in the comparison for the individual months. Figures 3.10 to 3.12 show that the simulation results of the model considering the seasonal difference reproduces the historical data better than the model which does not consider the seasonal difference.

3. Verification and validation of the typhoon model

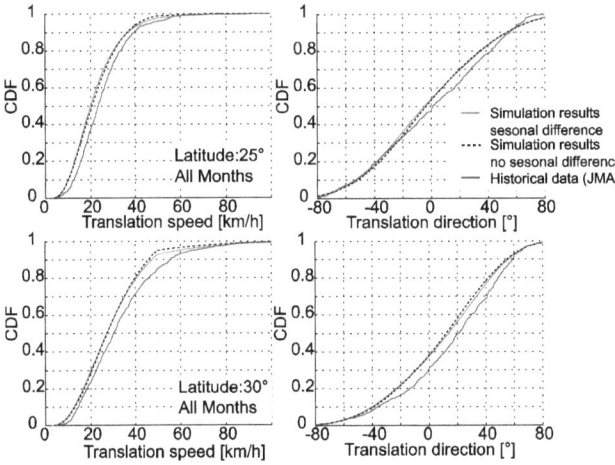

Figure 3.9: Comparison of the effect of considering the seasonal difference in the transition model for all months.

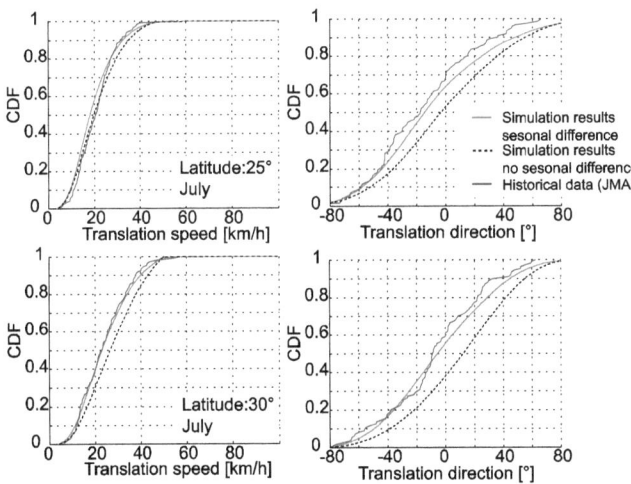

Figure 3.10: Comparison of the effect of considering the seasonal difference in the transition model for July.

3. Verification and validation of the typhoon model

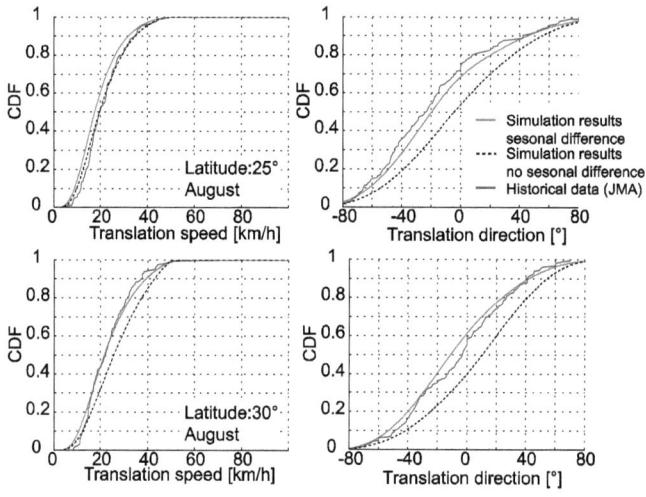

Figure 3.11: Comparison of the effect of considering the seasonal difference in the transition model for August.

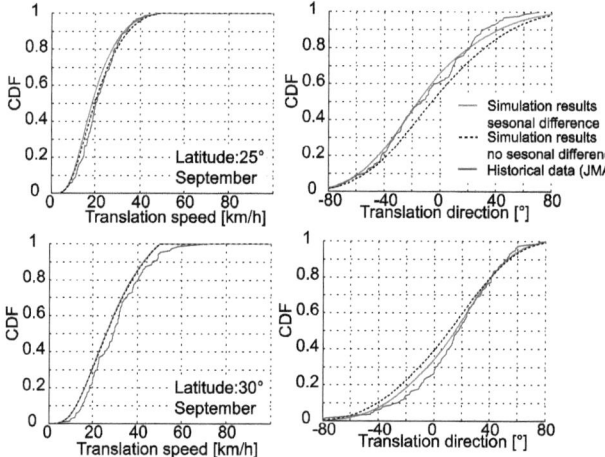

Figure 3.12: Comparison of the effect of considering the seasonal difference in the transition model for September.

3.3. Winds induced by typhoons

The performances of the developed wind field model and surface friction model are assessed by comparing the time histories of the observed wind speeds (10-minute sustained wind speeds) and the wind directions with those which are computed using the developed wind field model and surface friction model together with the JMA best track data as the input to these models. Figure 3.13 (left and centre) shows the time histories of the wind speeds and directions at two meteorological stations during the typhoon event Bart and Yancy. The computed wind speeds show good agreement with the observed wind speeds especially at the maximum wind speeds during the event. Figure 3.13 (right) shows the comparison of the maximum wind speeds (observed wind speeds vs. computed wind speeds) at all the JMA stations for the same typhoon event. Figure 3.14 shows the same comparisons for the typhoons Mireille and Songda.

Whereas the computed and observed wind speeds scatter, these are highly correlated and unbiased. One of the main reasons for the scattering of the computed and observed wind speeds can be possible failure to estimate "correct" roughness categories and corresponding roughness lengths of surrounding areas; different roughness categories/lengths result in large differences in the computation of the wind.

3. Verification and validation of the typhoon model

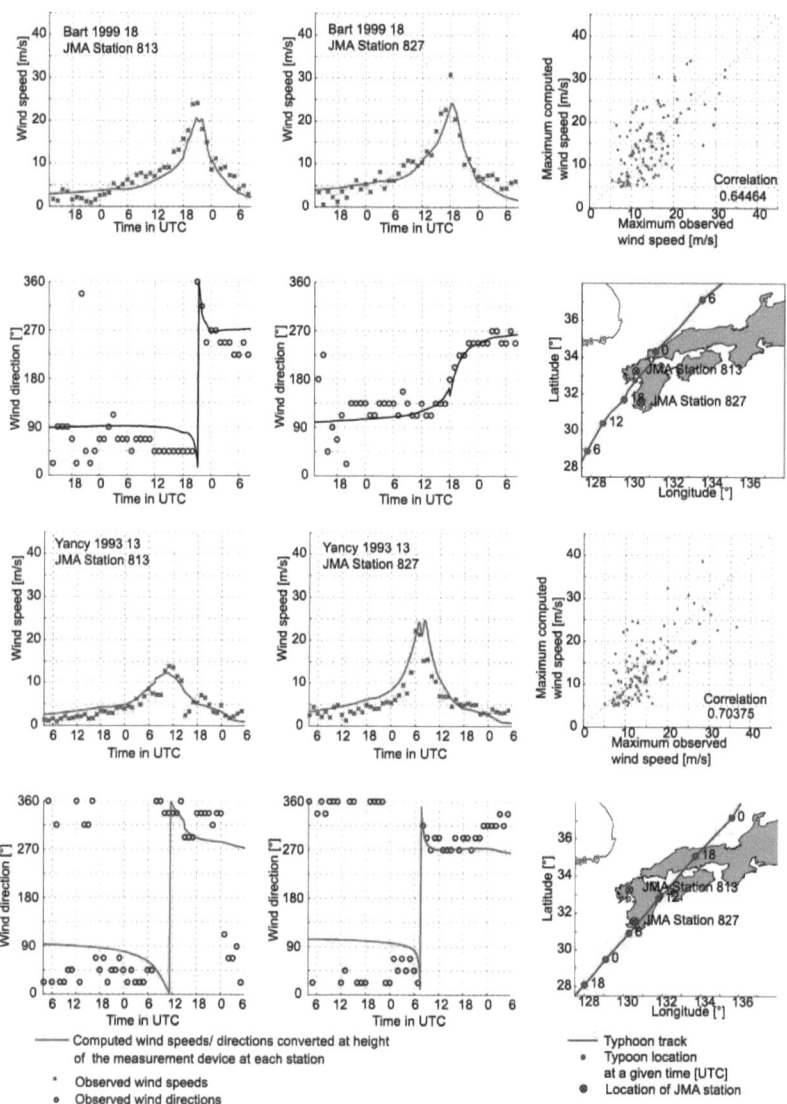

Figure 3.13: Time histories of the wind speeds, wind directions and the maximum wind speeds at two meteorological stations during several historical typhoon events.

3. Verification and validation of the typhoon model

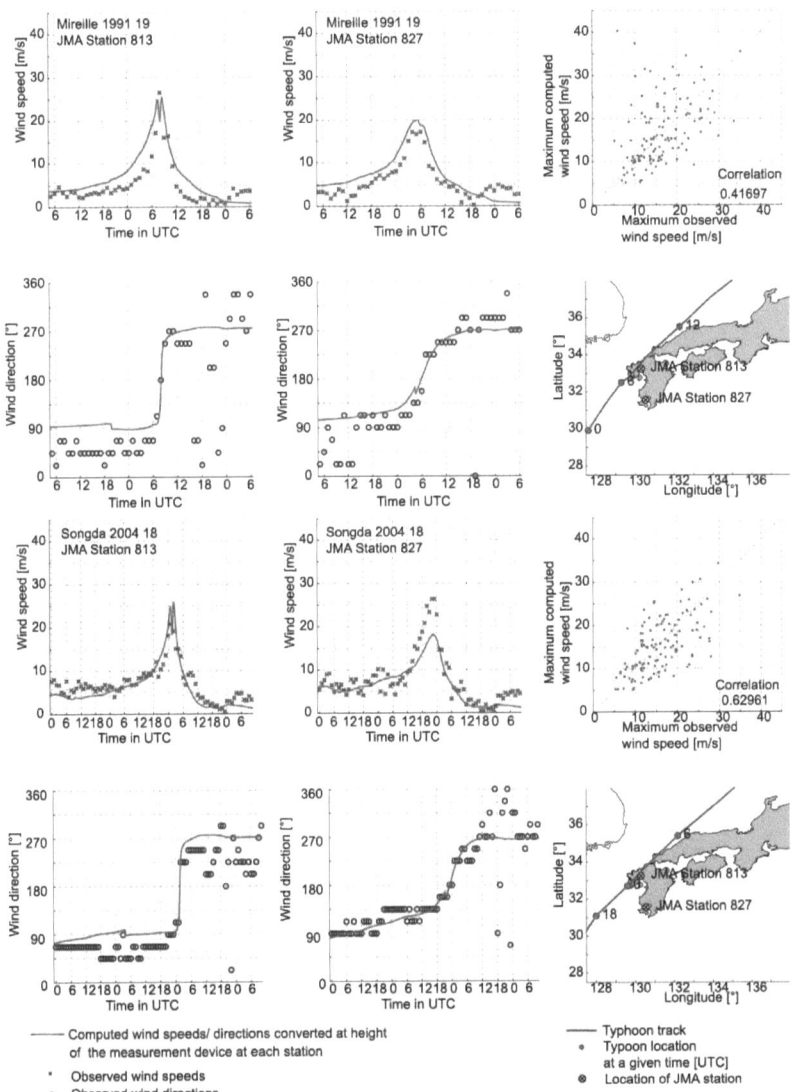

3.14: Time histories of the wind speeds, wind directions and the maximum wind speeds at two meteorological stations during several historical typhoon events.

3.4. Wind hazard map

For the overall verification of the hazard part of the developed typhoon model, a wind hazard map is established for the area in Japan using the hazard part of the typhoon model and is compared with the wind hazard map provided by the Architectural Institute of Japan, see (Architectural Institute of Japan, 2004) [10]. The comparison is made for the wind speeds corresponding to 100-year and 500-year return periods (hereafter, referred to as 100-year wind speed and 500-year wind speed respectively). Table 3-1 shows the comparison at several locations in Japan, see also Figure 3.15. The 100-year and 500-year wind speeds obtained using the developed hazard model are very close to those provided by the AIJ except for Okinawa[11]. However, it should be mentioned that there is a tendency that in the AIJ wind hazard map the differences of the wind speeds at coastal area and at neighboring inland area are more significant. The reason for this is not clear as the detailed procedure used by AIJ is not available. One possible reason may, however, be the different assumptions regarding the filling model. This should be investigated further.

---

10 The wind hazard map provided by the AIJ considers not only strong winds induced by typhoons but also due to other wind storm events, such as winter storms. However, it is reasonable to assume that the extreme events of strong winds, such as events corresponding to 100-year and 500-year return periods are caused by typhoons in most of the regions of Japan. Note also that whereas the development of the wind hazard map provided by the AIJ takes basis in the simulation using probabilistic models, the simulation is made not for the entire lifetime of typhoons; the typhoon events are simulated starting from the coast lines of the Japanese islands.

11 In the wind hazard map provided by the AIJ, the *maps* for 100-year and 500-year wind speeds are not provided for Okinawa. Instead, these are provided in terms of *table* without differentiating the locations within Okinawa (Note that Okinawa consists of many islands which locate far away from each other). Thus, it is not possible to precisely compare between the wind speeds obtained using the Aon-ETH model and the wind speeds provided by the AIJ for Okinawa.

3. Verification and validation of the typhoon model

Table 3-1: Comparison of 100-year and 500-year wind speeds.

Location	100-year wind speeds		500-year wind speeds	
	ETH	AIJ	ETH	AIJ
Hokkaido	30	32	35	36
Miyagi	32	32	36	36
Toyama	33	32	37	36
Chiba	34	36	38	38
Saitama	34	36	38	38
Kochi	35	37	40	41
Kumamoto	36	38	40	42
Okinawa	41	50	46	58

(Unit: m/s)
(*ETH refers to the hazard part of the developed typhoon model.)

Figure 3.15: Locations at which 100-year and 500-year wind speeds are compared in Table 3-1.

-91-

4. Treatment of epistemic uncertainties in the typhoon model

The consideration of the uncertainties involved in the modelling of the typhoons and in the portfolio risk analysis is a main focus of the developed Bayesian framework for probabilistic modelling of typhoon risks. The epistemic uncertainties in the developed typhoon model due to the modelling of the phenomena are quantified for each sub model. The statistical based occurrence model and the transition model directly consider the modelling uncertainties. The uncertainties associated with the deterministic wind field model and the surface friction model are considered indirectly in the vulnerability model (See Section 6.1.1). Apart from the statistical uncertainties involved in estimating the parameters of the sub models, epistemic uncertainties arise also from the selection of a model and due to the assumptions made in the typhoon model. For each sub model, several different model approaches are possible. Selecting one model approach, specifying the boundary conditions, choosing the appropriate data set and defining the assumptions in the model contributes to the epistemic uncertainties.

This chapter addresses the issue of the integration of the epistemic uncertainty into typhoon risk assessment. First, an approach for identifying and quantifying epistemic uncertainty is presented, based on the methodology proposed by Nishijima et al. (2011). On this basis, and restricting the focus to the epistemic uncertainty associated with the typhoon transition model, the variability of the analysis results due to the epistemic uncertainty is quantified. For example, the variability is quantified in terms of the relevant statistics such as annual probability of failure of a structure and portfolio losses corresponding to certain return periods.

In this study, the typhoon risk model developed in (Graf et al. (2009)) is employed as a basis, and the model is extended to encompass the epistemic uncertainty associated with the typhoon transition model. The approach presented here, however, is general, and can be applied to the other parts of the typhoon risk model as well as anonymous typhoon risk models. Also the epistemic uncertainties due to the assumptions made in the models should be investigated with the same procedure.

4.1. Introduction

Over the last years the standard methodology for the probabilistic modeling of typhoon events has been established. There, due to different assumptions in regard to the modeling of phenomena inherent in typhoon events and different sets of data employed in the development of the models, the hazard/risk analyses with different models usually result in different evaluations of the hazard/risk; however, this variability of the analysis results is not fully appreciated in practical typhoon risk assessments, although it is well recognized.

From a Bayesian statistical perspective, the variability of the models is understood as the consequence of the lack of knowledge and data, and is treated as epistemic uncertainty. Within the Bayesian statistical framework, the treatment of the epistemic uncertainty together with the aleatory uncertainty, which represents the randomness of phenomena in nature, has been discussed, and the rationale and mathematical formulation for the treatment of the epistemic uncertainty are presently readily available.

The following section presents a framework how the epistemic uncertainties due to the model selection can be quantified and integrated in the risk assessment and the decision analysis (Graf and Nishijima, 2011).

4.2. Background

During the last few decades the probabilistic modeling of natural hazards and their risks generally has experienced significant improvement and many successes not only in methodology, but also in their applications. Important applications include risk/reliability-based structural design, determination of design loads in building design codes, insurance loss estimations of engineered facilities and portfolios, and risk mapping to facilitate efficient resource allocations for risk reduction measures. More recently, research has been directed also to apply these probabilistic models to facilitate real-time decision making such as evacuation of people and assets in the face of emerging natural hazards, see e.g. Nishijima et al. (2009) and Anders and Nishijima (2011).

In parallel to this, a number of models for tropical cyclones have been developed, primarily in industrial domain, but also in public/academic domain (e.g. HAZUS (Department of Homeland Security, 2011) and FPHLM (Hamid et al., 2010)); see the IWTC-VII report (2010) for the overview of the state of the art for tropical cyclone risk modeling. It is fair

4. Treatment of epistemic uncertainties in the typhoon model

to say that presently decision makers concerning the management of tropical cyclone risks are readily accessible to these models.

Sharing the common methodologies and similar data sets in the modeling, however, models developed based on these often result in significantly different assessments of risks. These differences come from the use of different data sets, different modeling schemes, and different specifications of the modeling schemes. An example of the latter is the size of spatial/temporal discretization, in which the underlying random phenomena are considered to be homogenous. In spite of the presence of advanced statistical techniques, the identification of the best modeling scheme and the best specification are difficult, hence, these are often highly subjective, presumably contributing to the large variability of the risk assessments.

4.3. Status in practice and state-of-the-art

This type of variability is generally understood as epistemic uncertainty, which arises from the lack of sufficient data and/or knowledge. Note that in contrast to the epistemic uncertainty, randomness in nature is called aleatory uncertainty. The general treatment of both types of uncertainty in risk assessment and formal decision analysis has been, since decades, an issue of attention in civil engineering and other fields. Presently, the general framework for the treatment of the uncertainties is readily available, see e.g. Paté-Cornell (1996).

However, in practice the epistemic uncertainty is often ignored, mixed up with aleatory uncertainty otherwise treated in ad hoc manners, which can lead to erroneous assessment of risks. Such examples are investigated in detail in Nishijima et al. (2008b). Exceptions for these are found in the state-of-the-art projects for seismic hazard assessment for nuclear facilities at Yucca mountain in USA and in Switzerland (PEGASOS project). In these projects epistemic uncertainties are explicitly and consistently taken into account in the assessment of the seismic hazards. In regard to tropical cyclones, a unique study has been undertaken to quantify the variability of the risk assessment results using different models, taking basis in the areas of Florida and North Carolina in USA. (Watson and Johnson, 2004).

4.4. Challenging issues

The ultimate objective of probabilistic modeling and risk assessment is to facilitate decision makers to identify optimal decisions. Seen in this light, together with the current status described above, the following issues are addressed as future tasks in regard to the tropical cyclone risk management:

- Separation of aleatory and epistemic uncertainties.
- Quantification of epistemic uncertainty.
- Implementation of these uncertainties in the formal framework for risk assessment and decision analysis.

The explanation for the individual tasks is given in Section 4.5 with the introduction to the general framework for the uncertainty treatment.
As a first step for challenging these tasks, the present chapter focuses on the assessment of the hazard variability that comes from the use of different assumptions in the models, different modeling schemes and data

sets, and the specification of the modeling schemes; which is addressed as a part of the second task "quantification of epistemic uncertainty".

The present chapter is structured as follows. In Section 4.5 the general framework for the uncertainty treatment in risk assessment and decision analysis is introduced, whereby accounting for the tasks mentioned above. In Section 0, the typhoon risk model developed the author is introduced. This model is used as the reference typhoon model in Chapter 4, based on which several variants of the modeling schemes and the specification of the schemes are developed and risk assessments are systematically performed (Sections 4.7 to 4.10). The results are shown in Section 0. Discussion and conclusion follow.

4.5. General framework for uncertainty treatment

The general framework for the uncertainty treatment in risk assessment and decision analysis is briefly presented, taking basis in the review by Nishijima et al. (2008b).

A probabilistic modeling problem in risk assessment and decision analysis can in general be represented as a problem involving the expectation operation (in some cases a conditional expectation) over a function $g(\mathbf{X})$ of aleatory random variables $\mathbf{X} = (X_1, X_2, ..., X_n)$ as:

$$E[g(\mathbf{X})] = E_\Theta \left[E_\mathbf{X} [g(\mathbf{X}) | \Theta] \right] \qquad (4.1)$$

The random variables $\mathbf{X}$ are characterized by the joint probability distribution function $F_\mathbf{X}(\mathbf{x}|\theta)$ conditional on the epistemic random variables $\Theta = (\Theta_1, \Theta_2, ..., \Theta_m)$, which in turn are characterized by the probability distribution function $F_\Theta(\theta)$. Thus, $F_X(x|\theta)$ corresponds to the elicited probabilistic model and together with $F_\Theta(\theta)$ constitutes the probabilistic assessment model, see Figure 4.1. Here in the context of the present chapter, the aleatory random variables **X** represent the randomness on tropical cyclones and their consequences, and the epistemic random variables **Θ** represent possible modeling schemes and possible specification of the modeling schemes. Note that the expected value $E[g(\mathbf{X})]$ can correspond to e.g. the expected utility (loss) in case where $g(\mathbf{X})$ is the utility (loss) function; probability of failure in case where $g(\mathbf{X})$ is a an indicator function, which takes the value of one for the failure domain; otherwise zero.

Equation (4.1) and Figure 4.1 show the roles of aleatory and epistemic uncertainties in probabilistic modeling; and importantly, the order of the integration of the function $g(\mathbf{X})$ over the possible realizations of the set of the variables $(\mathbf{X}, \mathbf{\Theta})$. As shown in Nishijima et al. (2008b), a violence of this can lead to an erroneous assessment of the expected value of $g(\mathbf{X})$; e.g., probability of failure or risk. Here is the context where the first challenging issue is addressed.

Having separated aleatory and epistemic uncertainties, the epistemic uncertainty has to be quantified in terms of the probability distribution function $F_\Theta(\theta)$; hence, the second challenging task.

4. Treatment of epistemic uncertainties in the typhoon model

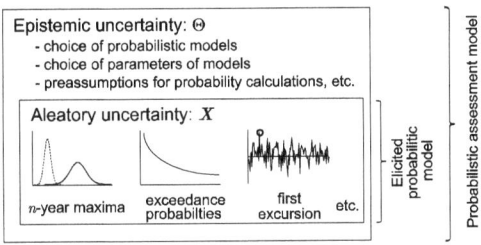

Figure 4.1: Probabilistic assessment subject to aleatory and epistemic uncertainties (after Nishijima et al. (2008b)).

Finally, the decision analysis is formulated by taking into account the decision variables in the framework. Here, two types of decision variables are distinguished; i.e. action decisions **a**, and test decisions **b**, see Jensen and Nielsen (2007). Action decisions change the characteristic of the underlying random phenomena. For instance, retrofit of a structure can change the structural performance of the structure. Test decisions change the decision makers' perception on the underlying the phenomena. For instance, consideration of detail information on land profile may reduce the uncertainty of the estimation of the wind speed induced by tropical cyclones. The objective function in the decision analysis is thus given as:

$$E[g(\mathbf{X},\mathbf{a},\mathbf{b})\,|\,\mathbf{a},\mathbf{b}] = E_\Theta\left[E_\mathbf{X}[g(\mathbf{X},\mathbf{a},\mathbf{b})\,|\,\Theta,\mathbf{a}]\,|\,\mathbf{b}\right] \qquad (4.2)$$

where the function g is now also a function of **a** and **b**. The optimal decisions are identified based on the objective function. The formulation of such decision problems is the third challenging task.

The proposed framework to integrate the epistemic uncertainties due to the different assumptions in the models, different model selection and the use of different data sets, in the decision analysis contains five steps as shown in Figure 4.2. First, all relevant alternative models have to be identified

4. Treatment of epistemic uncertainties in the typhoon model

and implemented. Second, the variability of the hazard assessment due to the different alternative models can be investigated. Third, to combine the different alternative models a weight, which represents the degree of believe for each alternative model has to be selected. Fourth, the combination of the weighted alternative models can be used to quantify the uncertainties due to the model selection. Fifth, the uncertainties due to the model selection have to be implemented into the framework for the risk assessment and decision analysis. In the following example step one and two are investigated and implemented into a example.

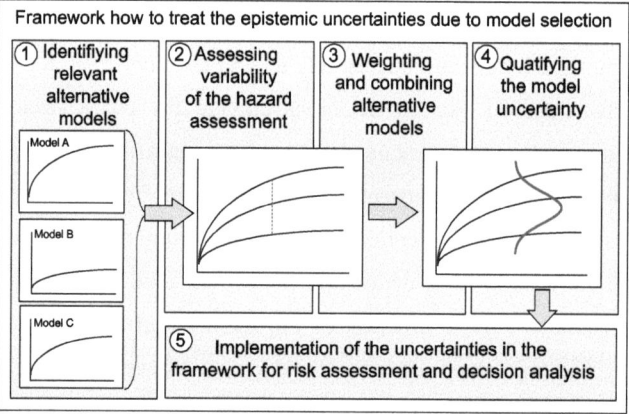

Figure 4.2: Framework to integrate epistemic uncertainties due to model selection into the decision analysis.

4.6. Reference Typhoon model

The typhoon model introduced in Chapter 2 is utilized as the reference typhoon model (Graf et al., 2009). Having briefly repeated the structure of the typhoon model in Section 4.7, the transition model (introduced in Section 2.6), which is a part of the typhoon model, is recapped in Section 4.8, since the variability due to the different modeling schemes for the transition model and their specifications are focused and investigated in the following.

4.7. Overview of the typhoon model

The developed typhoon model consists of two parts; a hazard model and a vulnerability model. The hazard model describes the probabilistic nature of the entire life-time of typhoons and associated wind fields from their occurrence to dissipation. The hazard model is composed of four sub-models; i.e. occurrence model, transition model, wind field model and surface friction model. The vulnerability model represents the probability distribution of the loss of individual exposures as a function of the wind speed, see Figure 4.3.

4. Treatment of epistemic uncertainties in the typhoon model

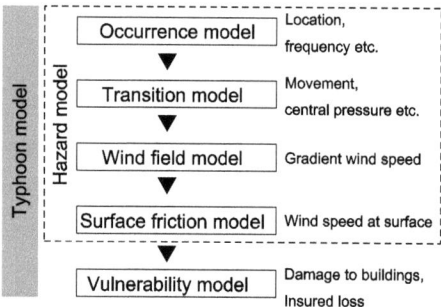

Figure 4.3: Components of the developed typhoon model.

4.8. Transition model

The transition model describes the probabilistic characteristics of the movement of typhoons and the change of the intensity of typhoons. The state of a typhoon is described by the following variables: the translation speed and direction, central pressure and radius of maximum wind speed of the typhoon. Therein, the spatial and temporal non-homogeneities of the probabilistic characteristics of these variables are considered. Namely, the probabilistic model for the translation is developed for each month (from January to December) and each 5°-by-5° grid (see Figure 4.4); the probabilistic model for the central pressure is developed for each month and each of the 18 different zones (see Figure 4.5).

4. Treatment of epistemic uncertainties in the typhoon model

Taking basis in the approach presented in Vickery et al. (2000) the changes of the state concerning the movement of typhoons are modeled by Markov chains (Brooks, 1998) as:

$$\Delta \ln V_i = a_1 + a_2 \ln V_i + a_3 \Gamma_i + \varepsilon_V \qquad (4.3)$$

$$\Delta \Gamma_i = b_1 + b_2 V_i + b_3 \Gamma_i + b_4 \Gamma_{i-1} + \varepsilon_\Gamma \qquad (4.4)$$

where V_i is the translation speed [km/hour] and Γ_i is the translation direction [°] at time step i. The time interval between the subsequent time steps is 6 hours. $\Delta \ln V_i$ and $\Delta \Gamma_i$ respectively represents the difference of the logarithms of the translation speeds and of the translation directions between the subsequent time steps. The coefficients (a_1, a_2, a_3) and (b_1, b_2, b_3, b_4) and the probability distributions of the random terms ε_V and ε_Γ are estimated for each grid, each month and each of easterly and westerly headed typhoons using the JMA best track data.

The variables concerning the intensity of typhoons are the central pressure and the radius of maximum wind speed. The change of the central pressure is modeled by:

$$\begin{aligned} P_{C,i+1} = c_1 + c_2 P_{C,i} + c_3 P_{C,i-1} + c_4 P_{C,i-2} \\ + c_5 T_i + c_6 \Delta T_i + \varepsilon_{P_C} \end{aligned} \qquad (4.5)$$

where $P_{C,i}$ is the central pressure [hPa] of a typhoon at time step i, T_i is the SST at the location where the typhoon is located at time step i and ΔT_i is the difference between the SSTs at the locations of the typhoon at time steps i and $i+1$. The coefficients $(c_1, c_2, ..., c_6)$ and the probability distribution of the random term ε_{P_c} are estimated for each zone and each month using the JMA best track data in the period between 1951 and 2006 and the SST data in the period between 1971 and 2000.

The model described above can be applied only when the typhoon is at sea; after the typhoon makes landfall, the filling model below is applied until the typhoon either dissipates or passes through the land. The filling model is represented (Vickery, 2005) as:

$$\Delta P_t = \Delta P_0 \cdot \exp(-(d_1 + d_2 \Delta P_0)t) \tag{4.6}$$

where ΔP_0 is the central pressure of the typhoon at the moment of the landfall, ΔP_t is the difference of the central pressure of the typhoon at time t and the peripheral pressure (here, 1013 [hPa] is assumed), and t is the time [hour] elapsed since the landfall. The coefficients (d_1, d_2) are estimated using the JMA best track data.

The radius of maximum wind speed is modeled as a random variable in the area around Japan, based on the historical data set.

4. Treatment of epistemic uncertainties in the typhoon model

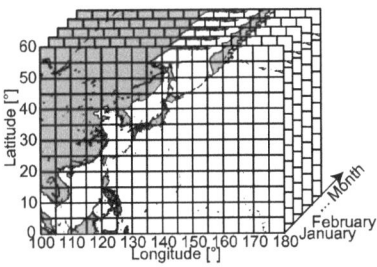

Figure 4.4: Spatial grids and temporal slices for the probabilistic model for translation.

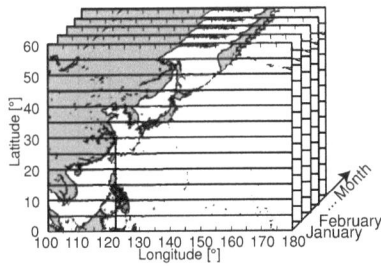

Figure 4.5: Spatial zones and temporal slices for the probabilistic model for central pressure.

4.9. Alternative models

To investigate the variability due to the different modeling schemes and their specifications in the transition model, eight different alternative models are established. The parameters considered here are: discretization in time and space, the order of Markov chains for the transition, and different data sets. The alternative models are labeled from 1 to 8, and the reference model introduced in the previous section is labeled as 0, see Table 4-1.

4.9.1. Discretization in space and time

The alternative models 1 to 3 represent different discretization in which the probabilistic characteristics are assumed homogeneous. The grid sizes of the respective models are $2°$-by-$2°$, $4°$-by-$4°$ and $10°$-by-$10°$. The alternative model 4 does not consider the seasonal difference of typhoon development in different months; i.e. the probabilistic characteristics of typhoons are identical in all the months.

Here, a tradeoff between the justification of the homogeneity in discretized time and space and the number of samples available becomes a critical issue. In general, a finer discretization is more likely to capture inhomogeneous characteristics, if any; however, the lack of sufficient number of samples may lead to a failure to judge the significance of the differences in the characteristics and also lead to unreliable models due to the large statistical uncertainty; as a consequence, the choice of the grid sizes are often subjective. Note here that in principle such a statistical uncertainty could be taken into account in the framework introduced in Section 4.5, which however is usually not the case in practice.

4.9.2. Functional form of transition

The alternative model 5 considers a different order of Markov model for the typhoon movement, which corresponds to the model proposed by Kerry Emanuel (2006b). This replaces the Equations (4.3) and (4.4) by the following two equations:

$$\Delta \ln V_i = a_1 + a_2 \ln V_i + \varepsilon_V \qquad (4.7)$$

$$\Delta \Gamma_i = b_1 + b_2 \Gamma_i + b_3 \Gamma_{i-1} + \varepsilon_\Gamma \qquad (4.8)$$

4.9.3. Functional form of intensity

The alternative model 6 replaces Equation (4.5) which concerns the change of the central pressure by:

$$P_{C,i+1} = c_1 + c_2 P_{C,i} + c_3 P_{C,i-1} + c_4 T_i + c_5 \Delta T_i + \varepsilon_{P_C} \qquad (4.9)$$

The alternative model 7, which is proposed by Jianming Yin (2009), considers the previous three time steps of the central pressure without considering the sea surface temperature (SST) as:

$$P_{C,i+1} = c_1 + c_2 P_{C,i} + c_3 P_{C,i-1} + c_4 P_{C,i-2} + \varepsilon_{P_C} \qquad (4.10)$$

4.9.4. Data sets

The best track data set provided by the China Meteorological Administration (hereafter, CMA) in the period between 1949 and 2008 is used to establish alternative model 8, instead of the best track data set provided by the JMA.

4.10. Overview of the alternative models

Table 4-1 shows the list of the alternative models and the differences of the parameters relative to the reference model. Each alternative model is systematically developed; hence, no subjective "tuning-up".

Table 4-1: Summary of the alternative models. The differences relative to the reference model 0 are highlighted.

Model	Equation used to estimate			Grid size	Data set	Seasonal difference
	$\Delta \ln V_i$	$\Delta \Gamma_i$	$P_{C,i+1}$			
0	(4.3)	(4.4)	(4.5)	5°x5°	JMA	yes
1	(4.3)	(4.4)	(4.5)	2°x2°	JMA	yes
2	(4.3)	(4.4)	(4.5)	4°x4°	JMA	yes
3	(4.3)	(4.4)	(4.5)	10°x10°	JMA	yes
4	(4.3)	(4.4)	(4.5)	5°x5°	JMA	no
5	(4.7)	(4.8)	(4.5)	5°x5°	JMA	yes
6	(4.3)	(4.4)	(4.9)	5°x5°	JMA	yes
7	(4.3)	(4.4)	Display	5°x5°	JMA	yes
8	(4.3)	(4.4)	(4.5)	5°x5°	CMA	yes

4.11. Variability of hazard assessment between alternative Models

Using these alternative models for the transition, and the other parts of the typhoon model developed by the author, Monte Carlo simulations are performed to investigate the variability of the hazard assessment.

4.11.1. Variation of the statistics on typhoon transition

Expected values of the annual numbers of typhoons that intersect the line segments at latitudes 30° and 35° within the range of the longitudes of 120° and 160° (see Figure 4.6) are first compared.

4. Treatment of epistemic uncertainties in the typhoon model

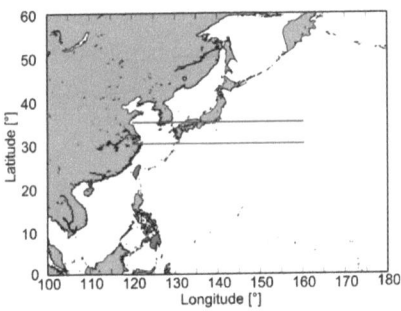

Figure 4.6: Line segments at which the numbers of typhoons are counted.

Figure 4.7 shows the expected values of the annual numbers for the individual alternative models and the estimates using the best track data sets. It is seen that the expected values do not significantly differ for the different alternative models. The expected values obtained by the simulations tend to be slightly lower than the estimates from the best track data sets. A possible reason for this is that the criteria differ for terminating the simulation of a typhoon and for terminating the record of a typhoon in the data sets.

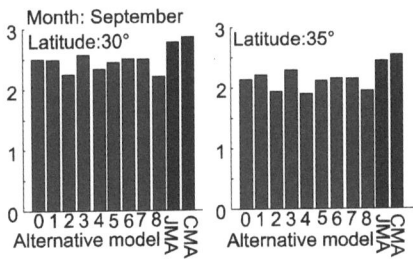

Figure 4.7: Expected numbers of typhoons that intersect the line segments at different latitudes.

-109-

4. Treatment of epistemic uncertainties in the typhoon model

Figure 4.8 and Figure 4.9 show the comparisons of the transition angle and speed as well as central pressure at the line segment at latitudes 30° and 35°. In Figure 4.10, the variability is illustrated for the individual parameters (i.e. discretization, order of Markov models, and data set).

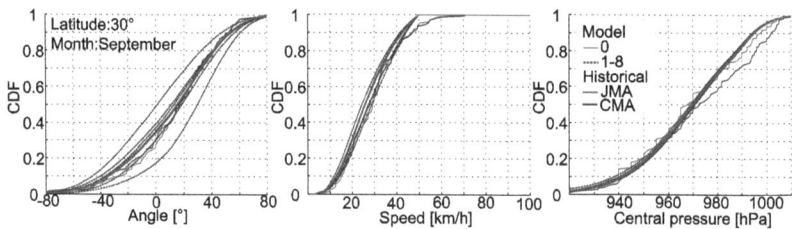

Figure 4.8: Cumulative distributions of the direction of movement (left), of the translation speed (centre) and of the central pressure of typhoons (right) crossing the latitude of 30° between the longitude [120,160°] in September for the different alternative models.

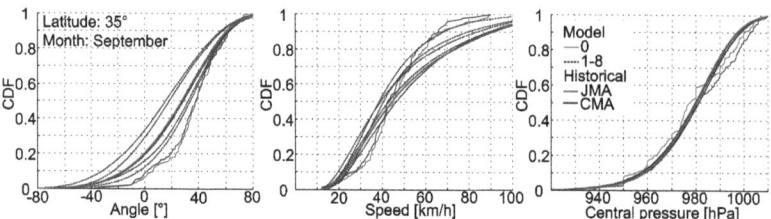

Figure 4.9: Cumulative distributions of the direction of movement (left), of the translation speed (centre) and of the central pressure (right) of typhoons crossing the latitude of 35° between the longitude [120,160°] in September for the different alternative models.

4. Treatment of epistemic uncertainties in the typhoon model

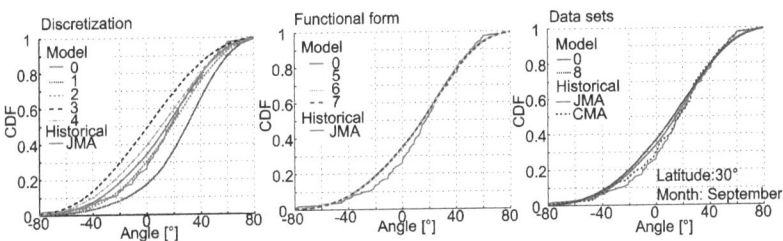

Figure 4.10: Cumulative distributions of the direction of movement of typhoons crossing the latitude of 30° between the longitude [120,160°] in September for the alternative models with different discretization (left), different functional form (centre) and different data set (right).

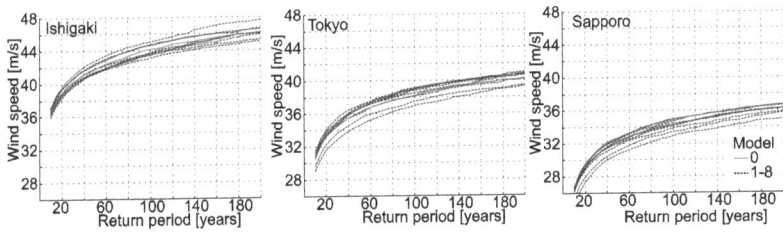

Figure 4.11: Maximum wind speeds as a function of return period for Ishigaki (left), Tokyo (centre) and Sapporo (right) for the different alternative models.

4. Treatment of epistemic uncertainties in the typhoon model

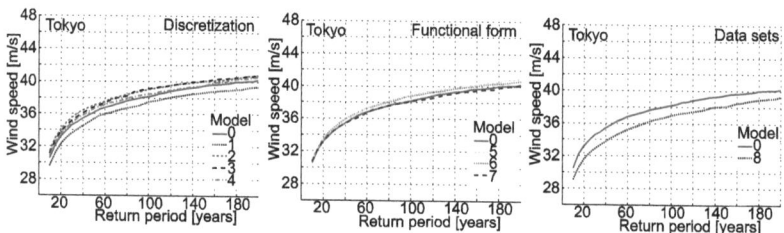

Figure 4.12: Maximum wind speeds as a function of return period for Tokyo for the alternative models with different discretization (left), different functional form (centre) and different data set (right). Variation of the maximum wind speed of the alternative models .

The maximum 10-minute sustained wind speeds for each typhoon event, adjusted to the height of 10 meters from the surface and roughness category II specified by the Architectural Institute of Japan (see AIJ load recommendation (Architectural Institute of Japan, 2004)), are considered. The wind speeds are simulated at three locations in Japan (Ishigaki, Tokyo and Sapporo, see Figure 4.13). Based on the simulations, the maximum wind speeds are estimated as a function of return period.

Figure 4.11 shows the maximum wind speeds at Tokyo, Sapporo and Ishigaki, which are obtained from the different alternative models. The maximum wind speeds obtained from the different alternative models differ by 2 to 3 [m/s] at the return period of 100 years. The variation increases at locations of higher latitudes. This can be explained by the fact that the difference in the state of typhoons is accumulated by the repetitive use of the different Markov models in the simulation. Figure 4.12 shows the variability associated with the individual variables at Tokyo. It is found that the major contribution to the variability is the discretization and data sets, while the contribution by the functional form is minor.

4. Treatment of epistemic uncertainties in the typhoon model

Figure 4.13: Locations of the cities where the annual maximum 10-min sustained wind speed are compared.

5. Updating of the typhoon model

Probabilistic models are typically implemented into risk management systems using whatever relevant information is available prior to the implementation. However, in the course of time more information becomes available and it is of significant practical importance to be able to update the probabilistic models based on the new information.

A main feature of the proposed Bayesian framework for probabilistic modelling of typhoon risk is the updating of the models with all the data available after one or more typhoon events have occurred. This feature facilitates to update the typhoon model after a certain period of time, for example in the end of a year, when all the information is organized as data. So, over time, the model will better represent the phenomena.

The following Sections 5.1 to 5.5 describe a theoretical approach as to how the new available information can be used in a most efficient way to update a hazard model.

For updating the developed typhoon model a model builder software tool has been created, which is described in Section 5.6. This model builder automatically establishes a typhoon model using all the available information as input.

5. Updating of the typhoon model

5.1. Background

The present chapter investigates a Bayesian approach for the updating of probabilistic models in the context of risk management of natural hazards. Bayesian probabilistic networks are proposed to form the basic tool for the probabilistic representation of knowledge and uncertainties. Updating of models is performed by instantiating the variables of the Bayesian probabilistic networks corresponding to observations from events of natural hazards. This approach, however, necessitates that large Bayesian probabilistic networks can be efficiently handled and for that purpose a compact object based representation of Bayesian probabilistic networks is suggested. The proposed methodology is applied to three illustrative examples considering updating of fragility model parameters. It is illustrated how commonly applied techniques for model updating in natural hazards risk assessments may lead to somewhat biased model parameter estimates. Furthermore, it is shown how available information on hazard intensities as well as on damages of structures can be utilized at the same time for the updating of the fragility model parameters in a consistent and efficient way.

5.2. Introduction

Probabilistic models play a central role in assessing and managing risks due to natural hazards. Typically a risk model is comprised by several individual model constituents, each of which is developed to model the relationship between input and output variables. Whereas the probabilistic models aim at describing the probabilistic characteristics of the constituents that are relevant to risks, the models themselves are in general subject to significant uncertainties. The main reason for this is on the one hand due to the natural variability of the hazard phenomenon itself (aleatory uncertainty) and on the other hand that the models are established as simplifications of the true nature based on imperfect engineering knowledge and/or limited data (epistemic uncertainty). Corresponding to the extent of simplifications and the imperfect understanding of the physical phenomena as well as to the degree of the availability of data, the models involve modelling uncertainties and statistical uncertainties. These types of uncertainties must be appropriately accounted for in the assessment of risks (Faber, 2005).The process of managing risks may be considered as a repeated sequence of actions with the purpose of optimizing measures of risk reduction, collection of information and updating of models. This perspective effectively implies that any probabilistic model should have the potential to incorporate available information in an efficient and consistent way. Herein, the term efficiency refers to the use of information available in such a way that the uncertainties of the quantities of interest can be reduced the most. The term consistency refers to the requirement that the developed probabilistic models should reflect the available knowledge and data in an unbiased manner and reflect the prevailing uncertainties and dependencies.

However, these requirements to probabilistic models are in fact often not fulfilled in practice; it is typical that information which might be obtained in regard to evolving hazard events is simply disregarded. Such information, however, if utilized for the purpose of updating, carry the potential of improving the accuracy of risk estimates. Furthermore, the data which can be collected after hazard events and provides a basis for improving the probabilistic models for representing both hazards and fragilities are sometimes utilized less than optimally and even incorrectly.

5.3. Problem setting

Probabilistic models used in risk management of natural hazards typically may be understood as individual constituents of the overall risk model. Constituents may e.g. be hazard models, fragility models and vulnerability models that represent the physical hazard phenomena, the resistance of structures exposed to hazards and the resulting losses in terms of e.g. financial costs, fatalities and environmental impacts. It is also possible that each constituent is composed of sub-models. These sub-models may be interconnected in such a way that the outputs from some models serve as the inputs to other models. In standard practice the constituent models are often established individually by experts representing different fields of engineering, natural and social sciences, on the basis of physical understanding, data and experience. Whereas this process in general is not problematic when risk models are established the situation is often different when models are updated e.g. on the basis of data which may be collected after hazard events.

In practical situations some of the variables in the probabilistic models are not observed directly or may not even be observable. In such cases the available information must first be transformed into the variables of interest before the updating can be performed. Consider as an example earthquake risk assessment. The peak ground accelerations (PGAs) at the locations where buildings are damaged are only seldom observed. In order to update the fragility models, however, the PGAs have to be estimated e.g. using seismic source data and the PGAs observed at seismological measurement stations together with the corresponding hazard models. The PGA geographical map obtained in this way is then applied to update fragility models with damage data from damaged structures, see e.g. (Basöz et al., 1999; O'Rourke et al., 2000; Shinozuka et al., 2000; Chen, 2003). The problem herein is that the uncertainties involved in the process of PGA estimations due to model and statistical uncertainties are often disregarded; the PGAs on the map are treated as if they were observed data. Consequently, the uncertainties introduced in this process are transferred into the fragility models, which in turn may result in biased estimates of the fragility models. The exception for this can be found in the papers by Straub and Der Kiureghian (2007) and Rossetto (2003) who explicitly take into account the model uncertainty associated with the empirical attenuation formula in the context of fragility model updating; this approach is also followed in the present chapter.

Typically applied approaches (in the forthcoming denoted "standard") for updating of probabilistic models in risk assessments whereby the constituent models are updated separately may result in an inefficient use of available information and data. For example, the observation that a larger number of buildings in a given area are damaged in a typhoon event as compared to other areas could be an indication that the maximum wind speed in this area was higher than in other areas. Thus, information about the observed damages and/or losses can be a source to estimate the maximum wind speed during a given event and is indeed useful for the purpose of reducing the uncertainties associated with the modeling of the maximum wind speed. Also, as soon as a hazard event has occurred the aleatory uncertainty becomes epistemic, since this uncertainty is no longer subject to a natural randomness but due to lack of knowledge. Measurements e.g. at meteorological stations during the hazard event can be utilized to reduce the former aleatory part of the uncertainty (see example 4.3). This leads to a smaller overall uncertainty in the estimation of the hazard index compared to the uncertainty in the hazard model.

5.4. Proposed approach

The diagram of risk management of natural hazards adopted in the proposed approach is shown in Figure 5.1. The diagram consists of two parts, i.e. "models of real world" and "real world." In the illustrated diagram three constituent models are considered; hazard model, fragility model and vulnerability model. These models can be represented by observable and unobservable variables. Indicators should be understood as any observable variables related to considered hazard events. The data obtained in the "real world" can be implemented into the models through the indicators.

5. Updating of the typhoon model

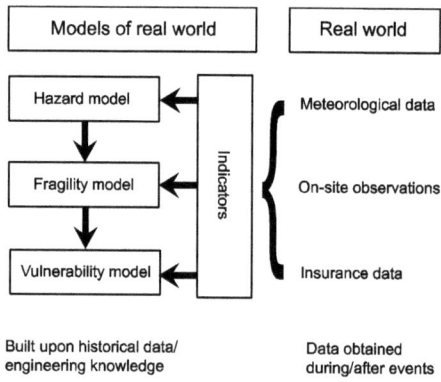

Figure 5.1: Diagram of risk management of natural hazards.

Here, a Bayesian approach is proposed for the general probabilistic modeling and for the purpose of updating. Bayesian probabilistic networks are utilized in the process of modeling, assessing and updating not only for matters of convenience but also for the purpose of communication; the graphical display of considered variables together with the causal relationships provide a strong means for bringing into the risk modeling the expertise of engineers and decision makers. All variables are represented in terms of nodes and dependencies through arrows; see Figure 5.2 (left). Figure 5.2 (left) shows how the constituent models may be represented using a Bayesian probabilistic network that corresponds to Figure 5.1. Without going into detail about the explanation of each node in the Bayesian probabilistic network (see detailed explanation in the following section), it should be mentioned that all three constituent models are interconnected through the input and output nodes. For instance, the hazard index w_i is the output from the hazard model and at the same time also the input to the fragility model, see Figure 5.2 (left). While each constituent model may be established separately using historical data and engineering knowledge, it is important that all constituent models should be interconnected and thus represented by an integrated Bayesian probabilistic network. This enables to update the models in a consistent and efficient way using the data that become available during/after hazard events. Figure 5.2 (center) shows how the data obtained in the "real world" can be used to update the models. In this figure two variables, i.e. H and F_1 are instantiated by the data (highlighted by "e"), where H represents the occurrence of a hazard event and F_1 represents the state of a structure (*failure* or *no failure*, for instance). By instantiating these two variables it is possible to e.g. update the probability density function of the variable D that represents the parameters of the fragility model or to calculate the conditional probability density function of W_1 etc. The advantage of the approach is that once all the constituent models have been established and been represented in an integrated Bayesian probabilistic network, it is straightforward to update the models using data that becomes available continuously during/after hazard events. For this purpose, a number of generic algorithms as well as software tools for Bayesian updating in the context of Bayesian probabilistic network representations are available e.g. (Lunn et al., 2000; Jensen, 2001; Gelman et al., 2004).

5. Updating of the typhoon model

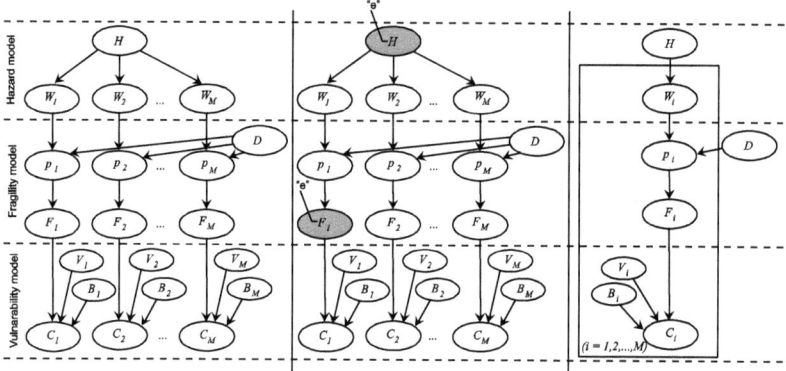

Figure 5.2: An illustration of a Bayesian probabilistic network (left) with instantiated nodes (center) and an equivalent abbreviated representation of the network (right).

An abbreviated object based representation may be useful to make the Bayesian probabilistic networks compact, see Figure 5.2 (right). Figure 5.2 (right) illustrates the abbreviated Bayesian probabilistic network which is equivalent to the networks in Figure 5.2 (left). In these Bayesian probabilistic networks the variables have identical marginal probability distributions and are conditionally independent given the common variables H and D. This abbreviation is useful in cases with many identical objects, e.g. when considering large portfolios or complex systems composed of multiple standard industrial components. In such cases, hierarchical Bayesian representations using object-oriented Bayesian probabilistic networks may also be useful, see e.g. (Kjaerulff, 1995; Gyftodimos and Flach, 2002) in general and (Nishijima and Faber, 2007) in particular for applications in civil engineering.

In the following section a Bayesian formulation is proposed and applied to three illustrative examples considering the updating of fragility models of structures subject to strong wind events caused by typhoons. Although the proposed approach can be utilized for the purpose of estimating e.g. portfolio losses due to emerging typhoon events using the information that becomes available from time to time (i.e. to calculate the conditional probability distribution of the portfolio losses given the present location and pressure of a given typhoon), this feature is not demonstrated in the following example, see e.g. (Faber, 2007) for further information. However, the proposed approach is readily applicable also for such situations.

5.5. Examples

In the following the proposed approach is illustrated through three examples. The focus of the examples is directed on the updating of a fragility model which represents the resistance of structures subject to strong winds induced by typhoons. The constituent models comprise a hazard model and a fragility model, see Figure 5.2; the vulnerability model in Figure 5.2 is not considered in the examples. The node H represents the occurrence of a typhoon event, which is a binary variable, i.e. *occurrence* or *no occurrence*. The node W_i represents the maximum wind speed at the ith location during a typhoon event, ($i = 1,2,...,M$). The node p_i represents the probability of failure of the structure at the ith location given the maximum wind speed W_i. The node Θ characterizes the fragility model, which commonly affects all nodes p_i; the structures considered in the examples are assumed to have identical probabilistic characteristics in regard to the resistance to strong wind. Finally, the node F_i represents the state of the structure at the ith location, i.e. *failure* or *survival*. The detailed probabilistic characteristics of these variables are given along with the examples.

The software tool WinBUGS (Lunn et al., 2000) is used for the Bayesian probabilistic network representation and for the updating of the models. WinBUGS is used because it can consider both continuous and discrete random variables and WinBUGS allows for directly representing the constituent models in a Bayesian probabilistic network. Other common software tools for Bayesian probabilistic networks like Hugin Experts (Hugin, 2006) or Genie (Genie, 2011) cannot handle continuous random variables. WinBUGS utilizes the Markov Chain Monte Carlo (MCMC) algorithm for the updating. Unlike the analytical or numerical updating of the probability distributions, the MCMC algorithm simulates the joint realizations from joint conditional probability distributions or joint posterior probability distributions, see e.g. (Gelman et al., 2004; Congdon, 2006; Gelman and Hill, 2007). The probabilistic characteristics of the probability distributions can be assessed from the realizations.

5.5.1.Proposed approach vs. standard approach for updating fragility models with data

This example investigates the extent to which the standard approach for updating of fragility models using data may result in a biased updating of the models. Furthermore, it is also shown that the proposed approach can update the probabilistic models appropriately. Here, the term standard approach refers to approaches characterized by the fact that the hazard

5. Updating of the typhoon model

index (maximum wind speed in this example) at the locations where the damage data are obtained are estimated from the hazard model before updating the fragility model and they are thus used for updating the fragility model disregarding the uncertainties involved in the estimation of these hazard indices.

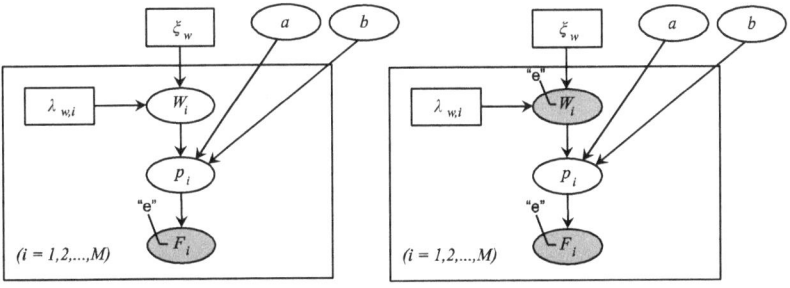

Figure 5.3: Considered Bayesian probabilistic network in Example 4.1 with instantiated nodes for the proposed approach (left) and for the standard and ideal approach (right).

For this purpose, the Bayesian probabilistic network shown in Figure 5.3 is considered. It is assumed that the hazard model is not associated with any statistical uncertainties, thus, is not to be updated with data; deterministic values are assumed for $\lambda_{w,i}$ and ξ_w that characterize the probability distribution of the maximum wind speed W_i. Given a typhoon event, W_i at the ith location ($i = 1, 2, ..., M$) are assumed to be statistically independent and to follow the lognormal distribution i.e.:

$$W_i \sim LN(\lambda_{w,i}, \xi_w) \tag{5.1}$$

where the parameter $\lambda_{w,i}$ is the mean of the logarithm of the maximum wind speed. $\lambda_{w,i}$ are defined by discrete values $\lambda_{w,i} = \ln(10 \cdot ((i-1) \bmod 25 + 1))$ and the number of samples M is a multiple of 25. ξ_w is the standard deviation of the logarithm of the maximum wind speed, which is assumed to be identical for all locations equal to 0.2. The distribution of W_i represents the aleatory uncertainty and depending on the load type an appropriate distribution function has to be selected. The fragility model is characterized by the probability of failure p_i, which in turn is characterized by two parameters a and b as:

$$p_i = \Phi\left(\frac{\ln w_i - a}{b}\right) \tag{5.2}$$

where $\Phi(\cdot)$ is the standard Normal cumulative distribution function and w_i is the maximum wind speed. Herein, the parameters a and b are assumed random variables; the fragility model is subject to statistical and model uncertainties. The updating of the fragility model is undertaken by updating the probability density functions of the parameters a and b. The probabilistic characteristics of the random variables are summarized in Table 1.

5. Updating of the typhoon model

Three different types of datasets for updating the fragility model are considered; the first dataset consists only of the damage states of the structures; the realizations $F_i = f_i$, ($i = 1, 2, ..., M$) are assumed known. The second dataset consists of the combinations of the states of structures and the maximum wind speeds at the location where the structures are located; i.e. $(W_i, F_i) = (e^{\lambda_i}, f_i)$, ($i = 1, 2, ..., M$) are known. Note that such a dataset corresponds to the situation where the maximum wind speed S_i is *pretended to be obtained* and are equal to the median of the maximum wind speeds that are derived from the hazard model. The last dataset consists of the combinations, i.e. $(W_i, F_i) = (w_i, f_i)$, ($i = 1, 2, ..., M$) are known.

The last dataset differs from the second dataset in the sense that the wind speeds are *really* obtained at the locations where the structures are located (it is, however, unlikely in practice). The first dataset (denoted "proposed" hereafter) represents realistic practical situations where the wind speeds are not measured at the locations where damage data on structures are obtained. Thus, the uncertainties of the maximum wind speeds must be considered in the process of updating (Figure 5.3 left). The second dataset (denoted "standard") represents the standard approach where it is assumed that the maximum wind speeds at the locations where the damage data are obtained are estimated from the hazard model and are represented by deterministic values (the median of the probability distribution function in this example) disregarding the uncertainties involved in the hazard model (Figure 5.3 right). The last dataset (denoted "ideal") represents an ideal situation where the maximum wind speeds measured at the locations where the damage data are obtained (Figure 5.3 right). The updating of the fragility model with the "ideal" dataset serves as a benchmark with the

5. Updating of the typhoon model

expectation that the rate of convergence of the parameters a and b is fastest in this case. These three different dataset are created by simulation according to the hazard model and the fragility models summarized in Table 5-1 and Figure 5.3. The true values of the parameters a and b are assumed to be equal to $\ln 100 \cong 4.605$ and 0.2 respectively; therefore, the parameters in the updated fragility models using the simulated datasets are expected to converge to these values as the sample size in the datasets increases.

Table 5-1: Parameters of the prior distribution of a and b.

Parameter	Distribution	Mean	Standard deviation
a	Normal	4.8	10
b	Gamma	20	8.9

Figure 5.4 and Figure 5.5 illustrate the rates of convergence for the parameters a and b respectively as function of the sample size for the three different datasets. The figures show the mean value of the posterior distribution of the parameters together with the 2.5% and 97.5% quantile values. The rate of convergence is fastest with the third dataset (ideal), whereas the rates of convergence with the other two datasets are almost identical. However, the updating with the second (standard) dataset fails to estimate the parameter b unbiased. The reason for this is that the disregarded uncertainties associated with the estimated wind speeds are transferred into the fragilty model, see Figure 5.6 (center) where the gradient of the curves is controlled by the parameter b and the smaller gradient corresponds to the larger model uncertainty. This is confirmed in Figure 5.6 (left) and (right) where $\xi_w = 0.01$ and $\xi_w = 0.5$ are assumed

5. Updating of the typhoon model

instead of $\xi_w = 0.2$. The larger the disregarded uncetainty of the estimated maximum wind speed, the larger the bias of the updated fragility model.

5.5.2. Updating of fragility models with the presence of common uncertainties

In the following the effect of the presence of common epistemic uncertainties in the hazard model on the updating of the fragility model is considered. The overall uncertainty is now composed of an aleatory part represented by the distribution of the wind speed W_i and of an epistemic part represented by ε^j introduced as a model uncertainty in the hazard model as:

$$V_i^j = \varepsilon^j \cdot W_i \qquad (5.3)$$

where ε^j represents the model uncertainty for the jth typhoon event, ($j = 1, 2, ...$) and follows a lognormal distribution with the mean of the logarithm being equal to $\lambda_\varepsilon = 0$ and the standard deviation of the logarithm being equal to ξ_ε. It is assumed that ε^j are identically and independently distributed. Thus, the maximum wind speed V_i^j follows a lognormal distribution with the parameters $\lambda_{V,i} = \lambda_{w,i} + \lambda_\varepsilon$ and $\xi_{V,i} = \sqrt{\xi_w^2 + \xi_\varepsilon^2}$ (Figure 5.7 left). In order to investigate the effect of the presence of the common model uncertainty, a parameter ρ is introduced as:

$$p = \frac{\xi_\varepsilon}{\sqrt{\xi_w^2 + \xi_\varepsilon^2}} \qquad (5.4)$$

which controls the ratio of the model uncertainty over the overall uncertainty involved in the hazard model. The effect of the presence of the common model uncertainty is investigated by changing the value of the parameter p but maintaining: $\sqrt{\xi_w^2 + \xi_\varepsilon^2} = 0.2$.

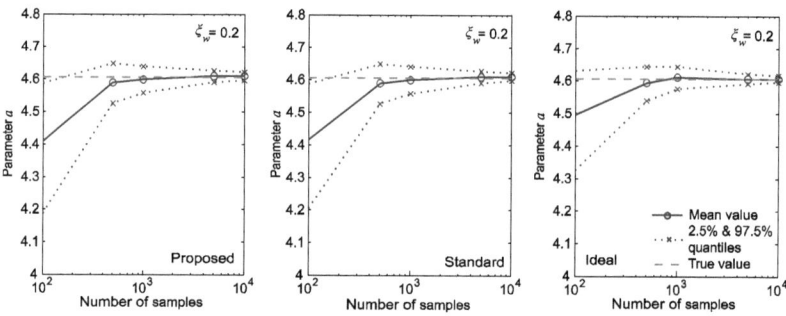

Figure 5.4: Rate of convergence of the parameter a as a function of the sample size used for the updating.

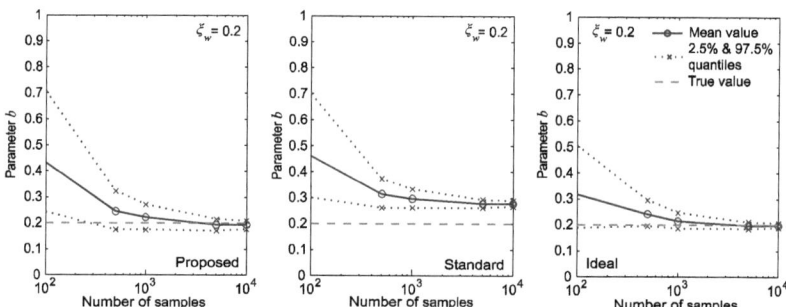

Figure 5.5: Rate of convergence of the parameter b as a function of the sample size used for the updating.

5. Updating of the typhoon model

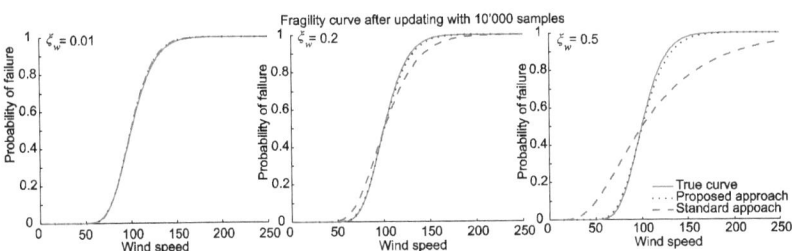

Figure 5.6: Fragility curve estimated by the different approaches for different ξ_w.

The datasets are created by simulation for different values of the parameter ρ. Other datasets are created with $\rho = 0.5$, but with a different number of typhoon events from which damage data are obtained. Each dataset contains 1000 samples and are utilized to update the fragility model which is characterized by the two parameters a and b as described in Example 5.5.1. The probabilistic characteristics of the variables are the same as in the Example 5.5.1. Figure 5.8 (left) shows that the updating of the fragility model provides biased parameters when the all data are obtained from one typhoon event and when the model uncertainty dominates. On the other hand, as the common model uncertainty decreases relative to the overall uncertainty and/or the number of typhoon event from which the data are sampled increases, the updating of the fragility model tends to be less biased and the parameters converge to the true parameters, see Figure 5.8 (right). This numerical investigation indicates that fragility models estimated using the dataset that consists only of data from one hazard event may be highly biased when the model uncertainty is dominating Thus, a key to update fragility models in practical situations is to reduce the model uncertainty; the model uncertainty can be readily reduced by conditioning the random variable ε that represents the model uncertainty using e.g. measured hazard indices at meteorological stations. An approach to conduct this is shown in the next example.

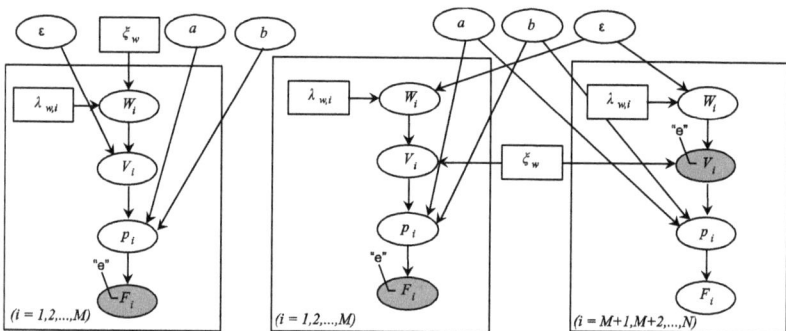

Figure 5.7: Bayesian probabilistic network including a model uncertainty ε for the proposed approach used in Example 5.5.2 (left) and Bayesian probabilistic network used in Example 5.5.3 (right).

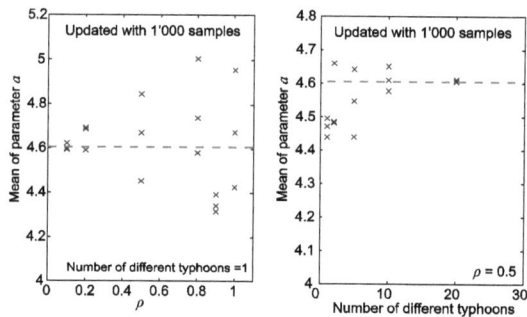

Figure 5.8: Mean value of the posterior distribution of parameter a after updating with datasets with different values of the parameter ρ (left) and with datasets with samples from different number of typhoon events (right).

5.5.3. Updating of fragility model using hazard intensities measured at meteorological stations

Maximum wind speeds measured at meteorological stations during the jth typhoon event can be used for conditioning the probability density function of the random variable ε^j as:

$$f(\varepsilon^j \mid v^j_{M+1}, v^j_{M+2}, ..., v^j_{M+N}) \tag{5.5}$$

where v^j_k, $(k = M+1, M+2, ..., M+N)$, are the measured maximum wind speeds at the meteorological stations during the jth typhoon event. Since the uncertainty becomes smaller by conditioning with these measurements, it is expected that the updating of the fragility models becomes more precise as is indicated in the previous example. Figure 5.7 (left) shows the corresponding Bayesian probabilistic network and Figure 5.7 (right) shows an equivalent Bayesian probabilistic network. The Bayesian probabilistic network shown in Figure 5.7 (right) is utilized for the convenience of updating. The equivalence of these Bayesian probabilistic networks is illustrated in the Appendix 10.3. The assumptions underlying the dataset simulations are the same as in Example 5.5.2, except that in the present example each dataset contains additional measured wind speeds v^j_k, to condition the model uncertainty ε^j, whereby the number of additional samples N is equal to the sample size M of the dataset which is used to update the fragility model.

Figure 5.9 illustrates the rates of the convergence of the parameters a and b respectively as a function of the sample size by conditioning the model uncertainty ε^j with N additional measured wind speeds v^j_k. As shown in Figure 5.9 it is possible to estimate the correct values of the parameters a and b even within one typhoon by using additional measured wind speeds v_k^j to reduce the model uncertainty ε^j.

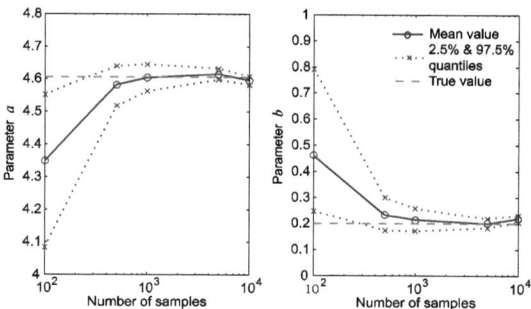

Figure 5.9: Rate of convergence of the parameter a (left) and b (right) as a function of the sample size used for updating.

5.6. Model builder software tool

To facilitate a user friendly way to update the typhoon model with new information, a model builder software tool has been developed. The model builder is written in matlab code and provides two parts one to update the occurrence model and the transition model and a second part to update the surface friction model.

The first part uses the best track data and the SST as input maps and automatically estimates all the parameters which are used for the occurrence model and the transition model. The typhoon model is updated by these parameters, estimated using the new available information.

5. Updating of the typhoon model

The second part uses the land use data and the topographic maps of the Japanese Islands as input and estimates the roughness length and the topological factor using the new information.

These two parts of the model builder enables the updating of the typhoon model in the required intervals as soon as new information is available.

The model builder is also used to establish a typhoon model a) which is based on only a part of the available data for the validation of the typhoon model as described in Chapter 3, b) which is based on a different functional form or data sets to assess the difference between alternative models as described in Section 4.9 and c) which is based on a data set obtained from a climate model to assess the effect of global warming as described in Chapter 7.

6. Application: Portfolio risk analysis

One of the main applications of the developed typhoon model is to estimate the statistics of insured portfolio losses in the insurance sector. This application shows how the Bayesian framework can be used to estimate portfolio losses due to typhoons with the consideration of the involved uncertainties. For estimating the statistics of insured portfolio losses, the hazard model described in Chapter 2 has to be extended with a vulnerability model, which describes the relation between the hazard index (namely the wind speed) and the losses. Section 6.1 describes the developed vulnerability model which is developed, Section 6.2 shows the verification of the vulnerability model and Section 6.3 explains how the vulnerability model is used to assess the portfolio risk.

Aon Benfield Japan provided historical damage observation data which helped to establish a data-based vulnerability model to estimate portfolio loss distributions. In order to fulfill the requirements of the confidentiality agreement between Aon Benfield Japan and ETH Zurich, the damage data and all results which are obtained by using the damage data are made anonymous and censored in this thesis.

To allow a user friendly risk assessment of a portfolio risk, a software tool with the name *TRAST* (Typhoon Risk Analysis Software Tool) is developed. *TRAST* provides a intuitive user interface to perform a risk analysis. It uses a database of a stochastic event set, which was created using the typhoon model described in Chapter 2 in combination with the

6. Application: Portfolio risk analysis

vulnerability model. The functionality of *TRAST* is explained in Section 6.4.

Due to the feature that the seasonal differences of the probabilistic characteristics of typhoon events are considered, it is possible to estimate portfolio losses in a certain period in a year. This is useful in practice when the assessments of portfolio losses are required for the remaining period of a year.

6.1. Vulnerability model

The vulnerability model describes the probabilistic characteristics of the ground-up loss[12] ratio of exposures as a function of the hazard index for different exposure types. The hazard index employed in the present typhoon model is the maximum 10-minute sustained wind speed at the nominal height during each typhoon event. The ground-up loss ratio Q is a random variable conditional on the hazard index w and the type of exposure[13] s, and is characterized by the probability density function $f_Q(q|w,\theta_s;s)$,

where θ_s is the set of the parameters which characterize the probability density function. In this chapter, the 10-minute sustained wind speed is referred to as wind speed and the ground-up loss ratio is referred to as loss ratio unless stated otherwise.

---

12 The insured losses of exposures are calculated based on the ground-losses and the policy conditions on the exposures.
13 A combination of the types of structure and object (building or content) of an exposure represents the type of the exposure.

6.1.1. Treatment of uncertainties

In order to take into account the uncertainties associated with the (deterministic) wind field model (Section 2.9) and surface friction model (Section 2.10) employed in the assessment of portfolio losses, whereby considering computational feasibility of the simulation of typhoon events, it is decided that these uncertainties are considered in the development of the vulnerability model indirectly; the uncertainties associated with the vulnerability model include not only the model uncertainties associated with the vulnerability itself but also the model uncertainties associated with the wind field model and the surface friction model.

The approach undertaken here for developing the vulnerability model is to create the hazard part of the datasets, which are utilized for the development of the vulnerability model, by using the wind field model and the surface friction model together with the best track data. Treating the wind speeds obtained in this way as if they were observed wind speeds in the estimation of the parameters of the vulnerability model, the uncertainties associated with the wind field model and the surface friction model are implicitly transferred to the uncertainties associated with the parameters of the vulnerability model. Notice that by using the same wind field model and surface friction model and the vulnerability model thus established in the assessment of portfolio losses the uncertainties associated with the wind field model and the surface friction model are considered. In the following sections the procedure for developing the vulnerability model in accordance with this approach is explained.

Note that in principle it is possible to separately consider the model uncertainties in the hazard model and the vulnerability model and to integrate consistently both uncertainties in the assessment of portfolio losses, see (Nishijima et al., 2009). However, this approach requires prohibitive computational efforts both in the estimation of the parameters of the hazard and vulnerability model and in the assessment of portfolio losses for such large portfolios as considered in this thesis. This is why the approach is not adopted in this thesis and the improvement of this is addressed as one of the tasks in future projects.

6.1.2. Exposure data and loss data available

The exposure data and loss data provided by the clients of Aon Benfield Japan through Aon Benfield Japan are utilized to develop the vulnerability model for the clients. The data of each client is utilized only for the purpose of developing the vulnerability model for the client and the data of the client is not utilized for developing the models for the other clients. In the following, the procedure for developing the vulnerability model is explained.

The loss data is provided for 20 relevant historical typhoon events occurred between 1998 an 2006. In this data, detailed information is available on each individual exposure to which payments were made in these typhoon events. The types of information available on each exposure and utilized in the development of the vulnerability model are:

- Amounts of the payments for damages of building and content
- Cause of damages, i.e. "wind storm" or "flood"[14]
- Insured values of building and content
- The location in terms of 7-digit post code
- Insurance policy
- Type of structure

Note that besides these types of information, other types of information are available in the data, such as occupancy type, construction year and number of stories of building; however these types of information are not utilized for the development of the vulnerability model; i.e. the developed vulnerability model is not differentiated by these indicators of the types of exposures.

---

[14] Specification of the cause of damages is made by a client of Aon Benfield Japan and it is possible that it does not correspond to the true cause of damages.

6. Application: Portfolio risk analysis

The detailed exposure data is provided as of 2005, 2006 and 2007. For other years, the summary of the exposures, i.e. the aggregated statistics on the exposures in each prefecture, is available. The detailed exposure data includes the following types of information for each individual exposure:

- Insured values of building and content
- The location in terms of 7-digit post code
- Insurance policy
- Type of structure

The vulnerability model is developed for each of following five types of structure:

- Wood
- RC
- Steel
- Block
- Unknown[15]

The geographical distributions of the exposures as of the years in which the 20 historical typhoon events occurred except for the year of 2005 are estimated using the summary of the exposures as of those years; i.e. assuming that the proportions of the numbers of exposures of given types of structure in given areas are constant over years, the numbers of

---

15 The type of structure "Unknown" may be utilized when the type of the structure of an exposure does not correspond to any of the other four types of structure. In fact, the vulnerability model for "Unknown" is developed using all the loss and exposure data without differentiating the types of the structures of the exposures; thus, the vulnerability model for "Unknown" is considered to correspond to the "average" vulnerability model weighted by the ratios of the numbers of exposures of different types of structure, which compose the portfolio. Note also that the vulnerability model for the type of structure "Stone", which is one of the type of structure indicated in the loss data, is not developed since the number of data for "Stone" is not sufficiently large for the statistical analyses. Thus, in the assessment of portfolio losses the type of structure "Stone" should be replaced by one of the five types of structure.

exposures for the types of structure in the areas as of those years can be estimated based on the geographical distribution of the exposure as of 2005 and the changes of the numbers of exposures. Note that as is explained in the next section the insured values of exposures in the exposure data are not required to develop the vulnerability model and thus need not to be estimated; i.e. the insured values of exposures in the loss data are sufficient to develop the model.

In the development of the vulnerability model, the *original* loss data, the *original* exposure data as of 2005 and the *estimated* exposure data for other years are utilized.

6.1.3. The flow of the development of the vulnerability model

The vulnerability model is developed in the following procedure, see also Figure 6.1. First, using the developed wind field model and surface friction model together with the best track data, the wind speeds are reproduced for each historical typhoon events for which the loss data is available. By doing so, it is possible to reproduce for all the historical typhoon events the wind speeds at the locations of all the exposures in the exposure and loss data. Then, two datasets are established. The first dataset (hereafter, loss dataset) contains for each exposure the following information:

- Reproduced wind speed
- Amounts of the payments for damages of building and content
- Cause of damages, i.e. wind storm or flood
- The location in terms of 7-digit post code
- Insured values of building and content
- Insurance policy
- Type of structure

Note that the cause of damages which is indicated in the loss data is changed from wind storm to flood if the reproduced wind speed is smaller than 11 [m/s]. In the development of the vulnerability model this is used as the indicator of the cause of damages. Note also that the information on the location and the insurance policy of each exposure is required to estimate the ground-up loss of the exposure.

6. Application: Portfolio risk analysis

The second dataset (hereafter, exposure dataset) contains the following information:

- Reproduced wind speed
- Insured values of building and content
- Type of structure

The first dataset is utilized to estimate the conditional probability distribution of the loss ratio for each type of exposure and cause of damages (wind storm and flood), given the occurrence of loss, as a function of the wind speed. Both the first and second datasets are utilized to estimate the probability of occurrence of loss for each type of exposure and cause of damages as a function of the wind speed.

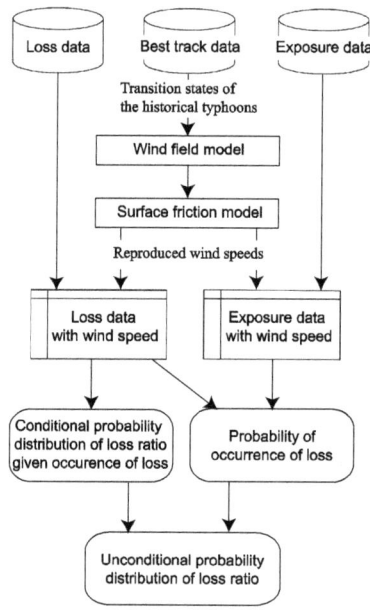

Figure 6.1: Flow of the development of the vulnerability model.

6.1.4. Conditional probability distribution of loss ratio given occurrence of loss

The conditional probability distribution $f_{Qloss}(q|w,\boldsymbol{\eta}_{s,c};s,c)$ for the loss ratio Q given the occurrence of loss is modeled for each respective type of structure s and cause of damages c to follow the log-normal distribution as:

$$f_{Qloss}(q|w,\boldsymbol{\eta}_{s,c};s,c) = \frac{1}{\sqrt{2\pi}\xi_{s,c}q}\exp\left(-\frac{(\ln q - \lambda_{s,c}(w))^2}{2\xi_{s,c}^2}\right) \quad (5.6)$$

where $\lambda_{s,c}(w)$ is the mean of the logarithm of the loss ratio, which is the function of the wind speed, i.e.:

$$\lambda_{s,c}(w) = \ln\left(a_{1,s,c} + a_{2,s,c}w\right) \quad (5.7)$$

and $\xi_{s,c}$ is the standard deviation of the logarithm of the loss ratio. Thus, the parameters to be estimated are $\boldsymbol{\eta}_{s,c} = (a_{1,s,c}, a_{2,s,c}, \xi_{s,c})^T$ for each type of exposure s and cause of damages c. Here, s and c respectively are the elements of the sets defined in the following:

$$s \in S = \left\{(s_{struct}, s_{obj}) | s_{struct} \in S_{struct}, s_{obj} \in S_{obj}\right\} \quad (5.8)$$

$$c \in C = \{\text{Wind storm, Flood}\} \quad (5.9)$$

where

$$S_{struct} = \{\text{Wood, RC, Steel, Block, Unknown}\} \quad (5.10)$$

-145-

$S_{obj} = \{\text{Building, Content}\}$ (5.11)

For the estimation of these parameters, first the loss ratios of the exposures for which payments were made are calculated using the information in the loss dataset. These are calculated as follows: the insured loss ratio of each exposure is calculated as the ratio of the payment over the insured value, and then by considering the policy condition for the insured loss ratio, the ground-up loss ratio is calculated. Then, employing the maximum likelihood method and using the dataset that consists of the combinations of the wind speed and the loss ratio of the exposures, the parameters $\eta_{s,c}$ are estimated. Note that the parameters $\eta_{s,c}$ are estimated for each type of exposure and cause of damages; thus, $5 \times 2 \times 2 = 20$ sets of the parameters are estimated. In Figure 6.2, the combinations of the loss ratio and the wind speed in the dataset, and the mean value and quantile values of the loss ratio calculated from the model (Equations (5.6) and (5.7)) with the estimated parameters $\eta_{s,c}$ are shown in the case of the exposure type (Wood, Building) and for both causes of damages (wind storm and flood). Note that as shown in the figure the minimum wind speeds above which the model for the loss ratio is established are 3.3 [m/s] for flood and 11 [m/s] for wind storm. This is because the minimum value of the wind speed calculated by the wind field model is 3.3 [m/s] (see Equation (2.10)) and the cause of damage is assumed to be flood if the wind speed is smaller than 11 [m/s]. The y-axis is omitted in Figure 6.2 to fulfill the confidentiality agreement between Aon Benfield Japan and ETH Zurich.

6. Application: Portfolio risk analysis

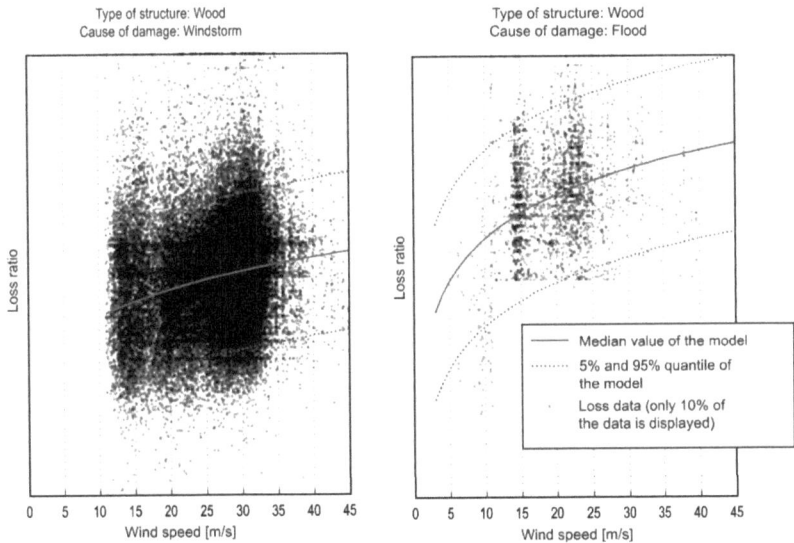

Figure 6.2: Loss ratio as a function of wind speed given the occurrence of loss (type of structure: Wood, types of object: building).

6.1.5. Probability of occurrence of loss

The probability $p(w|\varsigma_{s,c};s,c)$ of the occurrence of loss is modeled based on the logit model as:

$$p(w|\varsigma_{s,c},\varepsilon_{s,c};s,c) = \varepsilon_{s,c} \cdot \frac{1}{1+e^{-(b_{1,s,c}+b_{2,s,c}w)}} \qquad (5.12)$$

Here $\varepsilon_{s,c}$ represents the model uncertainties explained in Section 6.1.1. The random variable $\varepsilon_{s,c}$ is assumed to follow the log-normal distribution as:

6. Application: Portfolio risk analysis

$$f_{\varepsilon_{s,c}}(\varepsilon) = \frac{1}{\sqrt{2\pi}\psi_{s,c}\varepsilon} \exp\left(-\frac{(\ln\varepsilon - \omega_{s,c})^2}{2\psi_{s,c}^2}\right) \quad (5.13)$$

where $\omega_{s,c}$ is the mean of the logarithm of $\varepsilon_{s,c}$ and the $\psi_{s,c}$ is the standard deviation of the logarithm of $\varepsilon_{s,c}$. The parameters to be estimated are $\varsigma_{s,c} = (b_{1,s,c}, b_{2,s,c})^T$, $\omega_{s,c}$ and $\psi_{s,c}$.

For the estimation of these parameters, the wind speeds in the loss and exposure datasets are discretized by the interval of 1 [m/s]; thus it is possible to identify for each discretized wind speed the number of the exposures and the number of the exposures at which the loss occurs[16]. Then, employing the maximum likelihood method and using the dataset, which consists of the combinations of the wind speed, the number of the exposure and the number of the exposure at which the loss occurred, further setting $\varepsilon_{s,c} = 1$, the parameters $\varsigma_{s,c}$ are first estimated for each type of exposure and cause of damages. Therein, by trial and error it is decided that the model represented by Equation (5.12) is developed. As explained above, it is assumed that no loss due to wind storm occur below a wind speed of 11 [m/s]. Therefore the model representing the probability of occurrence of loss is developed for the case where the cause of damages is wind storm for the range of the wind speed $[11, \infty)$. Losses below 11 [m/s] are assumed to be caused by flood. The historical damage data indicates

---

[16] What is required for estimating the parameters without approximations is the indication on whether or not the loss occurred for each individual exposure in the exposure dataset. However, it is not possible to identify at which exposures in the exposure dataset the loss occurred by referencing the exposures listed in the loss dataset; however, it is possible to find these numbers (mentioned in the main sentence) by the discretization and these numbers are sufficient to estimate the parameters in an approximate way. Here, the approximation refers to the use of the discretized wind speeds.

that the damage data is composed by two different populations. This could may be explained by different failure mode of the buildings at a certain wind speed. Therefore were the parameters for the model estimated separately respectively for the ranges of the wind speed [11,26] and $(26,\infty)$ [m/s] in the case where the cause of damages is wind storm, and for the range of the wind speed $(3,\infty)$ [m/s] in the case where the cause of damages is flood. Whereby the parameters, the splitting point of the model (here 26 m/s) and the value below which no damage is assumed (here 11 m/s for the case of wind speed) were estimated using the maximum likelihood method.

The parameters $\varsigma_{s,c}$ are thus estimated for each range of the wind speed. Thereafter, the parameters $\omega_{s,c}$ and $\psi_{s,c}$ of the probability distribution of the random variable $\varepsilon_{s,c}$ are estimated using the dataset again and substituting the estimated parameters $\varsigma_{s,c}$ into Equation (5.12). The parameters $\omega_{s,c}$ and $\psi_{s,c}$ are common for all ranges of the wind speed.

The ratios of the occurrence of loss for the discretized wind speeds and the estimated probability of the occurrence of loss as a function of the wind speed in the case where the type of exposure is (building, wood) are shown in Figure 6.3. The y-axis is omitted in Figure 6.3 to fulfill the confidentiality agreement between Aon Benfield Japan and ETH Zurich. Note that the ratio shown in the figure is calculated as the ratio of the number of the exposures at which the loss occurred over the number of all the exposures for each wind speeds - implying that the *weight* of each point in the figure is not the same; however, in the estimations of the parameters the difference of the weights is taken into account. The estimated probabilities of the occurrence of loss are drawn by a solid line and dotted lines. The solid line corresponds to the mean of the probability and the dotted lines correspond to the probabilities with different values of the variable $\varepsilon_{s,c}$ (these values are the discretized values in such a way that each of the values has the equal mass probability of 10%.).

The above described approach can lead to a discontinuity of the model at the splitting point of the model (at 26 [m/s]) as can be seen in the Figure 6.3 (left). This could may be explained by different failure mode of the buildings at a certain wind speed. Note that Aon Benfield Japan provided two different loss data sets, a older and a newer more detailed data set. The functional form of vulnerability model was developed based on the first (older) data set. Appling the described model approach lead to a continuous model over the whole range of wind speeds. After receiving the second (newer) data set, the vulnerability model was re-established using the new data, which is shown in Figure 6.3. It was decided not to change the model approach and to accept the discontinuity at 26 [m/s] in order to be able to compare the results of the analysis, based on the different data sets.

6. Application: Portfolio risk analysis

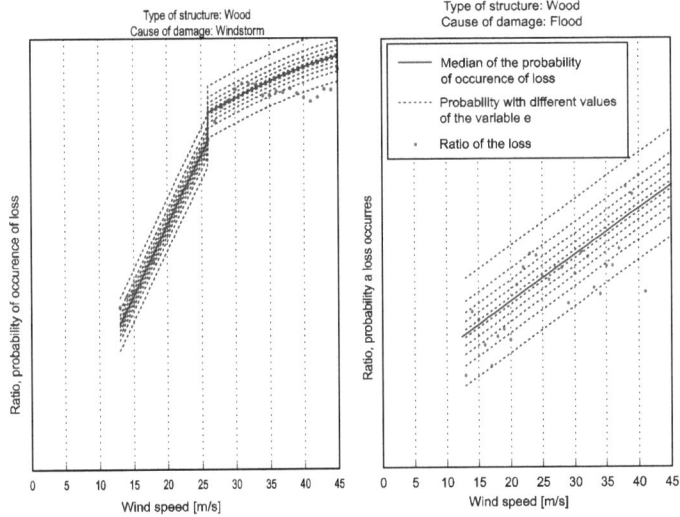

Figure 6.3: Ratios of the occurrence of loss and the estimated probabilities of the occurrence of loss.

6.1.6. Unconditional probability distribution of loss ratio

The unconditional loss distribution of loss ratio for each type of exposure is obtained using the probabilistic models obtained in the last two sections as:

$$f_Q(q\mid w,\boldsymbol{\theta}_s;s) = \begin{cases} \sum_{c\in C} f_{Q\mid loss}(q\mid w,\hat{\boldsymbol{\eta}}_{s,c};s,c)p(w\mid\hat{\varsigma}_{s,c},\varepsilon_{s,c};s,c) & (q>0) \\ \delta(q)\cdot\sum_{c\in C}\left(1-p(w\mid\hat{\varsigma}_{s,c},\varepsilon_{s,c};s,c)\right) & (q=0) \end{cases} \quad (5.14)$$

where " ^ " signifies the estimates of the parameters and $\boldsymbol{\theta}_s = (\varepsilon_{s,wind\,storm},\varepsilon_{s,flood})^T$ and $\delta(\cdot)$ is the delta function.

6. Application: Portfolio risk analysis

For illustrative purposes, *median-median* ground-up loss ratios are shown in Figure 6.4 for exposures whose type of object is "building" (for the five types of structures) as a function of the wind speed calculated using the developed vulnerability model. The curves $y_s(w)$ drawn in the figure correspond to the cases where the median values are taken for both the conditional loss ratios ($\hat{a}_{1,s,c} + \hat{a}_{2,s,c} w$) and the probability of the occurrence of loss ($\varepsilon_{s,c} = 1$), i.e.:

$$y_s(w) = \sum_{c \in C} \left(\hat{a}_{1,s,c} + \hat{a}_{2,s,c} w \right) \cdot \frac{1}{1 + e^{-(b_{1,s,c} + b_{2,s,c} w)}} \qquad (5.15)$$

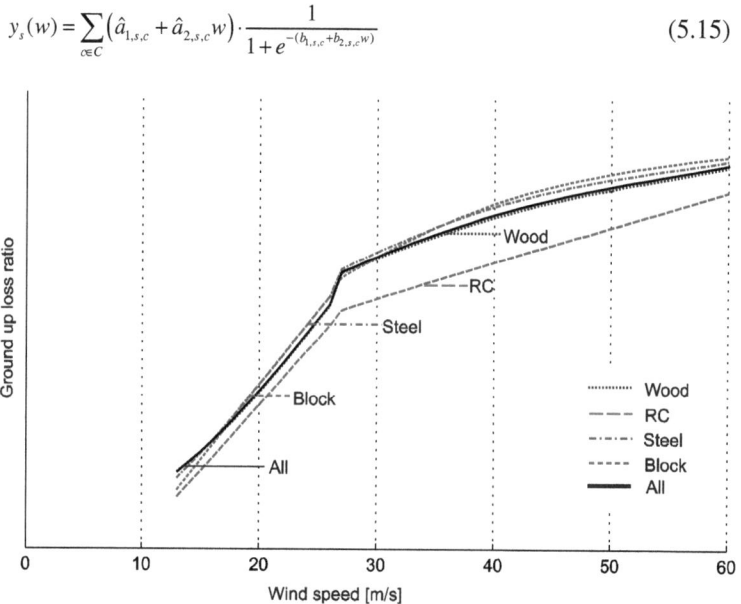

Figure 6.4: *Median-median* ground-up loss ratios for exposures whose type of object is "building".

The y-axis is omitted in Figure 6.4 to fulfill the confidentiality agreement between Aon Benfield Japan and ETH Zurich. Notice that the unconditional probability distributions thus obtained cannot directly be applied for the cases where the franchises are smaller than those which have been employed for the exposures included in the historical loss data. This is because: minor damages whose monetary losses do not reach the franchises are not recorded in the loss data. In order to apply the vulnerability model for such cases, it is required that engineering approaches for modelling such minor damages are employed and modified the developed (statistics-based) vulnerability model accordingly.

6.1.7. Dependency of the random variables $\varepsilon_{s,c}$

The parameters of the random variables $\varepsilon_{s,c}$ are estimated separately for the individual types of exposure and causes of damages; thus, no probabilistic structures of the dependency between the variables are considered. However, in the assessment of portfolio losses the assumptions are required on the dependency of the random variables $\varepsilon_{s,c}$ between different type of exposure as well as causes of damages.

It is assumed that for the dependency of the random variables $\varepsilon_{s,c}$:

- Full correlation between different causes of damages (wind storm and flood) for each type of exposure.

- Full correlation between different types of object (building and content) for each type of structure.
- Independency between different types of structure.

- Further investigation on the dependency is addressed as one of the future tasks.

6. Application: Portfolio risk analysis

6.2. Insured losses caused by historical typhoons

The insured losses are reproduced using the developed typhoon model for the relevant historical typhoon events for which the loss data is available. The insured losses are reproduced using the historical event set (see Section 6.3.2) with the estimated exposure data (see Section 6.1.2) for client of Aon Benfield Japan. Figure 6.5 show the comparisons of the insured losses available in the loss data and the *mean* reproduced insured losses for the relevant historical typhoons. The reproduced losses well represent the historical losses for strong as also for weak typhoon events. The y-axis is omitted in Figure 6.5 to fulfill the confidentiality agreement between Aon Benfield Japan and ETH Zurich.

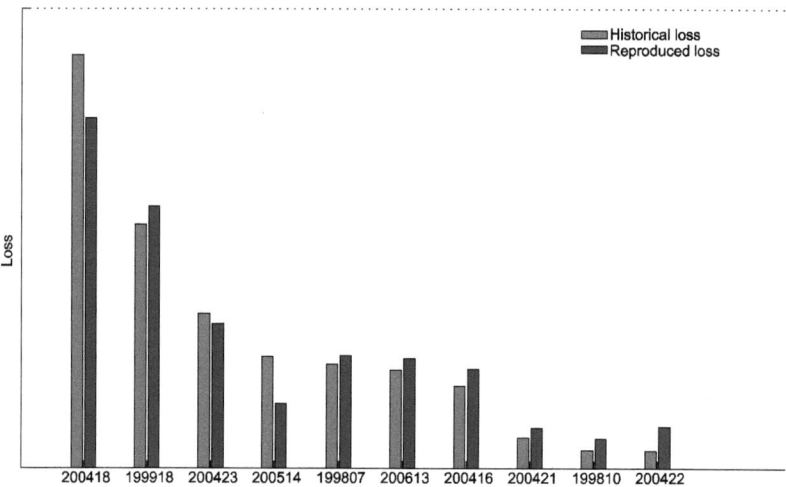

Figure 6.5: Comparison of historical losses and reproduced losses using the developed typhoon model.

6.3. Assessment of typhoon risks

6.3.1. Mathematical formulation for assessment

Probability distribution of annual maximum losses

Let X denote a random variable representing the annual maximum insured loss (hereafter, loss refers to insured loss unless otherwise stated) of a given portfolio, and let $X_{i,j}$ denote random variables representing the loss of the portfolio due to the jth typhoon ($j = 0, 1, 2,...$) in month i, where $i = 1, 2, ..., 12$ correspond to January, February,..., December respectively. For convenience, it is assumed that the realization of $X_{i,0}$ is $x_{i,0} = 0$; this corresponds to the case where no typhoon occurs in a certain month i. The annual maximum loss X is represented using $X_{i,j}$ as:

$$X = \max_{i=1,2,...12} \max_{j=0,...,J_i} X_{i,j} \quad (5.16)$$

where J_i is the number of occurrences of typhoons in month i, which is a random variable. The conditional cumulative distribution function $F_{X|\Theta}(x|\theta)$ of the annual loss X given $\Theta = \theta$ is first to be assessed by:

$$F_{X|\Theta}(x|\theta) = P[X \leq x | \Theta = \theta] = E\left[I[X \leq x] | \Theta = \theta\right] \quad (5.17)$$

where $I[\cdot]$ is an indicator function that returns one when the condition in the bracket is satisfied and zero otherwise and the vector Θ consists of the model uncertainties for all types of exposure and causes of damages, i.e.:

6. Application: Portfolio risk analysis

$$\Theta = (\Theta_{Wood}, \Theta_{RC}, \Theta_{Steel}, \Theta_{Block}, \Theta_{Unknown})^T \quad (5.18)$$

The expectation term of Equation (5.17) is reformulated as:

$$E\left[I[X \leq x] | \Theta = \theta\right] = E\left[I\left[\left(\max_{i=1,2,\ldots 12} \max_{j=0,\ldots,J_i} X_{i,j}\right) \leq x\right] \middle| \Theta = \theta\right]$$

$$= E\left[E\left[I\left[\left(\max_{i=1,2,\ldots 12} \max_{j=0,\ldots,J_i} X_{i,j}\right) \leq x\right] \middle| J_1, J_2, \ldots, J_{12}\right] \middle| \Theta = \theta\right] \quad (5.19)$$

The inner expectation is with respect to $X_{i,j}$ ($i = 1, 2, \ldots, 12$ and $j = 1, 2, \ldots, J_i$) conditional on the numbers J_i of occurrences of typhoons. The outer expectation is with respect to the numbers J_i of the occurrences of typhoons in each month.

Taking basis in the Monte Carlo simulation technique, the conditional expectation given $\Theta = \theta$ can estimated by:

$$\hat{F}_{X|\Theta}(x|\theta) \approx \frac{1}{M} \sum_{m=1}^{M} I\left[\left(\max_{i=1,2,\ldots 12} \max_{j=0,\ldots,j_i^m} x_{i,j}^m(\theta)\right) \leq x\right] \quad (5.20)$$

where $x_{i,j}^m(\theta)$ is the realization of the portfolio loss for the simulated transition $\mathbf{z}_{i,j}^m$ of the jth typhoon in month i in the mth realization $\mathbf{j}^m = (j_1^m, j_2^m, \ldots, j_{12}^m)^T$ of $\mathbf{J} = (J_1, J_2, \ldots, J_{12})^T$ conditional on $\Theta = \theta$. M is the number of the simulations of one-year occurrences of typhoon events. The realizations of the occurrence and initial states of the typhoon are simulated by resampling from the historical data. The transitions $\mathbf{z}_{i,j}^m$ of the typhoons are simulated probabilistically using the transition model. In

6. Application: Portfolio risk analysis

the simulation of the transitions of the typhoons the mean value of the SST at each location on the Northwest Pacific Ocean for each month estimated from the SST data is employed. For each transition $z_{i,j}^m$, the maximum 10-minute sustained wind speeds $w_{i,j}^m$ at the nominal height are calculated at all 1km-by-1km grids on the Japanese islands using the wind field model and the surface friction model; thus, each exposure of the portfolio is required to locate to one of the grids at which the wind speeds are calculated. The way for this is explained in Section 6.3.3. Finally, the portfolio loss $x_{i,j}^m(\theta)$ is obtained using the vulnerability model conditional on $\Theta = \theta$, which is explained in the subsequent section.

Insured loss of exposures for given typhoon events

Let $g(q;c_{pol})$ represent the relation between the ground-up loss ratio q and the insured loss ratio for a given policy condition c_{pol} of an exposure. Using the vulnerability model $f_Q(q|w,\theta_s;s)$ and thus the relation $g(q;c_{pol})$, the mean value $\mu_{Q_{insured}}(w|\theta_s;s)$ of the insured loss ratio of the exposure can be calculated as:

$$\mu_{Q_{insured}}(w|\theta_s;s) = \int g(q;c_{pol}) f_Q(q|w,\theta_s;s) dq \qquad (5.21)$$

Note that often the insured loss ratio is also a function of the insured value of the exposure as is shown in Figure 6.6 and the insured value is schematically included in the policy condition c_{pol} in the formulation above.

In the assessment of portfolio losses, it is assumed that the insured loss of the portfolio in each typhoon event converges to its mean value. Notice, however, the insured loss is still a random variable because the mean value is a function of the random variables Θ, which represent the model uncertainties. The expected value of the insured loss of an exposure is obtained by multiplying the insured value with the insured loss ratio.

Finally, the portfolio loss $x_{i,j}^m(\Theta)$ for a given typhoon event and for a given θ is calculated as:

$$x_{i,j}^m(\Theta) = \sum_{n=1}^{N_{exposure}} \mu_{Q_{insured}}(^nw_{i,j}^m | ^n\theta_{n_s}; ^ns) \cdot {^nv_{insured}} \tag{5.22}$$

where $N_{exposure}$ is the number of exposures in the considered portfolio, n is the index of each individual exposure in the portfolio, $^nw_{i,j}^m$ is the wind speed at the location of the nth exposure (which corresponds to one of the elements of $\mathbf{w}_{i,j}^m$), ^ns is the type of the nth exposure and $^nv_{insured}$ is the insured value of the nth exposure.

6. Application: Portfolio risk analysis

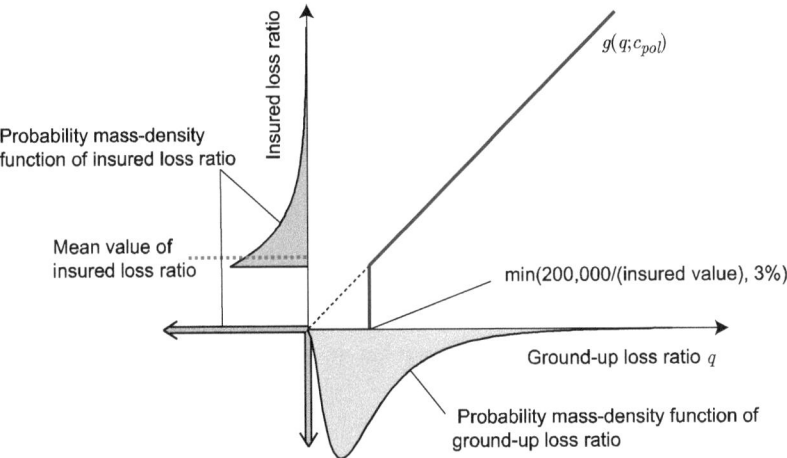

Figure 6.6: Example of the relation between ground-up loss ratio and insured loss ratio for a given wind speed.

Statistics of insured losses

The statistics of insured losses that are assessed in the software tool developed during this thesis are summarized in Table 6-1[17]. In what follows, the procedure for assessing these statistics is explained. Note, however, that other statistics can be assessed using the developed typhoon model and formulation described in the previous two sections.

For given model uncertainties θ, a conditional cumulative distribution $\hat{F}_{X|\Theta}(x|\theta)$ of the annual maximum loss of a portfolio is assessed, see Equation (5.20). The unconditional cumulative distribution $F_X(x)$ is assessed as:

---

[17] The capability of the assessment of these statistics in the software tool is requested by Aon Benfield Japan.

$$\hat{F}_X(x) = \int \hat{F}_{X|\Theta}(x|\theta) f_\Theta(\theta) d\theta \qquad (5.23)$$

where $f_\Theta(\theta)$ is the joint probability density function of the random variables Θ. By discretizing the joint probability density function, a joint probability mass function $p_\Theta(\theta)$ is obtained. Considering the assumptions on the dependency of the variables (see Section 6.1.7) and discretizing the continuous states of each variable into ten states with the equal probability, i.e. 10^{-1}, the joint states of the variables Θ are discretized into 10^5 states with the equal probability, i.e. 10^{-5}. Thus, Equation (5.23) can be approximated as:

$$\hat{F}_X(x) = \sum_{\theta \in \Omega} \hat{F}_{X|\Theta}(x|\theta) p_\Theta(\theta) \qquad (5.24)$$

where Ω represents the set that is composed of all the discretized states of the variables Θ and $p_\Theta(\theta) = 10^{-5}$ for all $\theta \in \Omega$. Based on Equation (5.24) it is possible to assess the losses for different return periods.

The mean value $\mu_{i,j}^m$ of the loss for each typhoon event can be assessed as:

$$\hat{\mu}_{i,j}^m = \sum_{\theta \in \Omega} x_{i,j}^m(\theta) p_\Theta(\theta) \qquad (5.25)$$

The standard deviation $\sigma_{i,j}^m$ of the loss for each typhoon event can be assessed as:

$$\hat{\sigma}_{i,j}^m = \sqrt{\sum_{\theta \in \Omega} (x_{i,j}^m(\theta) - \hat{\mu}_{i,j}^m)^2 p_\Theta(\theta)} \qquad (5.26)$$

6. Application: Portfolio risk analysis

The cumulative distribution function $F_{X,i,j}^m(x)$ of the loss for each typhoon event can be assessed as:

$$\hat{F}_{X,i,j}^m(x) = \sum_{\theta \in \Omega} I\left[x_{i,j}^m(\theta) < x\right] p_\Theta(\theta) \qquad (5.27)$$

The quantile vales can be assessed based on Equation (5.27).

Whereas the statistics mentioned above concern the maximum of the losses in a given year, the annual average loss (AAL) concerns the sum of the losses in a given year. Thus, Equation (5.16) must be replaced by:

$$X = \sum_{i=1}^{12} \sum_{j=0}^{J_i} X_{i,j} \qquad (5.16)'$$

Then, following the same procedure explained in the Section 6.3.1 *Probability distribution of annual maximum losses*, the expected value μ_{sum} of the sum of the losses in a given year (i.e. AAL) is assessed as:

$$\hat{\mu}_{sum} = \frac{1}{M} \sum_{\theta \in \Omega} \sum_{m=1}^{M} x_{i,j}^m(\theta) p(\theta) \qquad (5.28)$$

The distribution of the loss due to an emerging typhoon event is assessed as follows. First, using the transition model, the wind field model and the surface friction model together with the information available for the typhoon event as the initial condition, possible future typhoon tracks and corresponding wind fields are simulated. These are substituted as $\mathbf{z}_{i,j}^m$ and $\mathbf{w}_{i,j}^m$ respectively in the formulation in the Section 6.3.1 *Probability distribution of annual maximum losses*. Here, the subscripts i and j do not make any sense thus can be abbreviated as $\mathbf{z}^m$ and $\mathbf{w}^m$. Then, the loss

6. Application: Portfolio risk analysis

$x^m(\theta)$ for each possible track and corresponding wind field is calculated as a function of the variables θ in the same manner as explained in the Section 6.3.1. Finally, the cumulative distribution function $F_{X,cond}(x)$ of the loss due to the emerging typhoon event is assessed as:

$$\hat{F}_{X,cond}(x) = \sum_{\theta \in \Omega} \sum_{m=1}^{M} I\left[x^m(\theta) \le x\right] p(\theta) \qquad (5.29)$$

Table 6-1: Statistics assessed in the software tool *TRAST*.

Statistics	Utilized formula
Losses for different return periods (5,10,25,70,100,250, 500 and 1000 years)	Equation (5.24)
Annual average loss	Equation (5.28)
Mean value of the loss for each typhoon event	Equation (5.25)*
Standard deviation of the loss for each typhoon event	Equation (5.26)*
5% and 95% quantiles of the loss for each typhoon event	Equation (5.27)*
Distribution of the loss due to an emerging typhoon event	Equation (5.29)

(* In the case of historical typhoon events, the subscripts i and j and the superscript m do not make any sense; $z_{i,j}^m$, $w_{i,j}^m$ and $x_{i,j}^m(\theta)$ should be interpreted as the transition, wind field and loss of the considered historical typhoon event respectively.)

Combining two portfolio loss assessments

In the case where the losses of two portfolios are already assessed individually and the results are stored, the statistics of the loss of the portfolio which consists of the two portfolios can be assessed without performing the calculation shown in the previous sections. This is possible if the insured losses $x_{i,j}^m(\theta)$ are stored for each typhoon event (i, j and m) and each realization of the model uncertainty Θ for two portfolios A and B, which are denoted as $_A x_{i,j}^m(\theta)$ and $_B x_{i,j}^m(\theta)$ respectively; i.e. the insured loss $_C x_{i,j}^m(\theta)$ of the combined portfolio C for each typhoon event (i, j and m) and each realization of the model uncertainty Θ is calculated as:

$$_C x_{i,j}^m(\theta) = {_A x_{i,j}^m(\theta)} + {_B x_{i,j}^m(\theta)} \tag{5.30}$$

The statistics of the insured loss of the portfolio C can be obtained by replacing $x_{i,j}^m(\theta)$ in Equation (5.20) with $_C x_{i,j}^m(\theta)$. Note that the information required for combining two portfolio loss assessments in this way is only the summed insured losses of two portfolios, i.e. $_A x_{i,j}^m(\theta)$ and $_B x_{i,j}^m(\theta)$; the information of individual exposures is not required.

This is useful in many situations including the following cases:

- Loss of a portfolio which consists of a large number of exposures can be assessed by splitting the portfolio into two or more mutually exclusive portfolios, assessing the losses of these portfolios individually (possibly with some PCs in parallel) and combining these loss assessments.

- Both loss of a portfolio and losses of parts of the portfolio (possibly separated according to the policy conditions) can be assessed efficiently by assessing the losses of parts of the portfolio and then combining these loss assessments to obtain the loss of the whole portfolio; i.e. additional calculation for obtaining the loss of the whole portfolio is not required.

6.3.2. Typhoon event sets

The assessment of portfolio losses by the Monte Carlo simulation technique requires a set of the realizations of typhoon events using the hazard model. Since the wind field model and the surface friction model utilized for the loss assessment are deterministic, the event set can be defined in terms of the transition $\mathbf{z}_{i,j}^m$ of typhoons or the wind field $\mathbf{w}_{i,j}^m$ induced by the typhoons ($i = 1, 2..., 12$, $j = 1, 2, ..., J_i$ and $m = 1, 2, ..., M$). The developed software tool stores the typhoon event set in both terms. The transitions $\mathbf{z}_{i,j}^m$ are utilized for visualizing the track of the typhoons. The wind fields $\mathbf{w}_{i,j}^m$ are utilized for assessing the portfolio losses as well as for visualizing the wind fields of the typhoons. In the present version of the software tool, the number M of years in the simulation of typhoon events is 23,992 and this amounts to 172,957 relevant typhoon events. All these typhoon events are stored in the software tool and these are utilized for the assessment of portfolio losses. This event set is called the stochastic event set.

The other event set, called historical event set, is also included in the software tool. The historical typhoon events included in the historical event set are the typhoon events from 1977 to 2006 in which the typhoons approached the Japanese islands and the event of the typhoon Vera (1959). In these typhoon events, the track records in the best track data are utilized to represent the transitions of the typhoons. Based on these the wind fields are reproduced and stored in the event set. The number of the typhoon events stored in the historical typhoon event set is 123.

6.3.3. Disaggregation of portfolios

The disaggregation of portfolios is required in cases where the exposure data are provided only in an aggregated manner. Here, the aggregated exposure data refer to the data in which the information on individual exposure is merged together and only the summary of the parameters of the aggregated exposures is available. For instance, an aggregated exposure data may contain only the number, the sum of the insured building values, the sum of the insured content values of exposures for each type of structure and policy situated in each Sompocode[18]. Hereafter, these quantities are denoted the parameters of the exposure.

---

18. Sompocode is the code utilized in the insurance industries in Japan for the identification of regions in which exposures are situated. The typical size of a region identified by a Sompocode is slightly larger than that of a ward.

6. Application: Portfolio risk analysis

The disaggregation of aggregated exposure data is made in two steps. The first step is to disaggregate the values of the parameters of aggregated exposures into those of individual exposures. The second step is to locate, applying a certain rule (explained later), each individual exposure to one of the 1km-by-1km grids at which wind speeds are calculated and stored in the event sets.

The assumption made in the first step is that the values of the parameters of the individual exposures which constitute the aggregated exposures are identical. Hence, the values of the parameters of the individual exposures are obtained by dividing the values of the corresponding parameters of the aggregated exposures by the number of exposures included in the aggregated exposures. For instance, when the sum of the insured building values of the exposures in a Sompocode is 104,810,400 [yen] and the number of exposures in the Sompocode is 12, the insured building value of each (disaggregated) individual exposure is obtained as 104,810,400/12=8,734,200 [yen].

In the second step the disaggregated individual exposures are located at one of the grids using an indicator I that is considered to be correlated to the geographical distribution of certain types of occupancy of exposures. For instance, "residential house" is one of the possible types of occupancy and the geographical distribution of residential houses is considered to be correlated to the geographical distribution of the households. In this case, the indicator I can be selected as the number of households. Note that the value of the indicator I must be available at ChoOhaza[19] level. Using the value of the indicator I at each ChoOhaza, the individual

---
19. ChoOhaza is a class of the geographical hierarchy commonly utilized in Japan, which corresponds to a community level.

exposures are located to specific ChoOhazas in proportion to the values of the indicator I at the ChoOhazas. Then, the individual exposures located in each ChoOhaza are distributed geographically uniformly to the 1km-by-1km grids that belong to the ChoOhaza. This rule is applied when the exposure data are aggregated at the level of prefecture, ward and Sompocode. In the case where the exposure data includes the information of the 7-digit post codes for identifying the locations of the exposures, the individual exposures are relocated to the grids closest to the locations of the exposures defined by the 7-digit post code. Therein, if the exposure data are aggregated, the first step explained above is applied in order to disaggregate the values of the parameters of the exposures. The diagram of the disaggregation is shown in Figure 6.7.

In Figure 6.8 an example of the disaggregation is shown for the area of Tokyo for the type of occupancy "residential house". Note that the datasets required for the disaggregation are provided by Aon Benfield Japan. These include:

- Geographical distribution of different type of the use of buildings, e.g. commercial office, residential house, factory etc.
- GIS data which defines the 1km-by-1km grid system
- Data that contains the correspondence between different hierarchical levels of locations, i.e., ChoOhaza, 7-digit post code, Sompocode, ward and prefecture

6. Application: Portfolio risk analysis

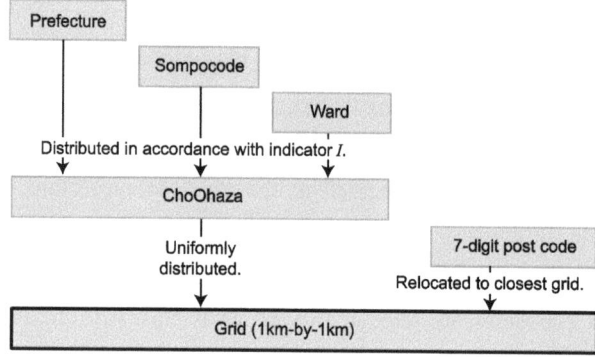

Figure 6.7: Diagram on the disaggregation of aggregated exposure data.

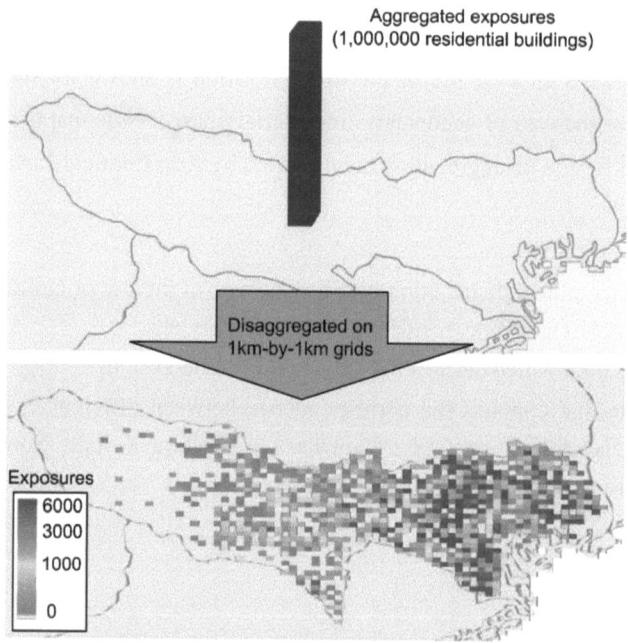

Figure 6.8: Example of the disaggregation in the area of Tokyo.

6.3.4. Program components

The program developed in this thesis is composed of three parts: hazard event set builder, vulnerability model builder and the main program, see Figure 6.9. The hazard event set builder is required for creating the stochastic event set. The vulnerability model builder is required for developing and updating the vulnerability model. For the assessment of portfolio losses in practice, only the main program is required. The main program is called *TRAST (Typhoon Risk Analysis Software Tool)* and is provided as a stand-alone software tool with graphical user interface, the main functions of the software tool *TRAST* is explained in Section 6.4.

The hazard event set builder and the vulnerability event set builder are basically written in MATLAB (Version: 2008b). However, for the calculations related to the Bayesian probabilistic networks and the Bayesian statistics, WinBUGS[20] and Hugin[21] are utilized. It is requested by Aon Benfield Japan that the graphical user interface of the main program should be written in Visual Basic, thus Visual Basic.NET 2005 is utilized as the platform of the program development. The codes of the calculation of portfolio losses are written in MATLAB 2008b and these codes are complied to dynamic link libraries (DLLs) using MATLAB compiler and included in the main program.

---

20 See http://www.mrc-bsu.cam.ac.uk/bugs/winbugs/contents.shtml.
21 See http://www.hugin.com.

6. Application: Portfolio risk analysis

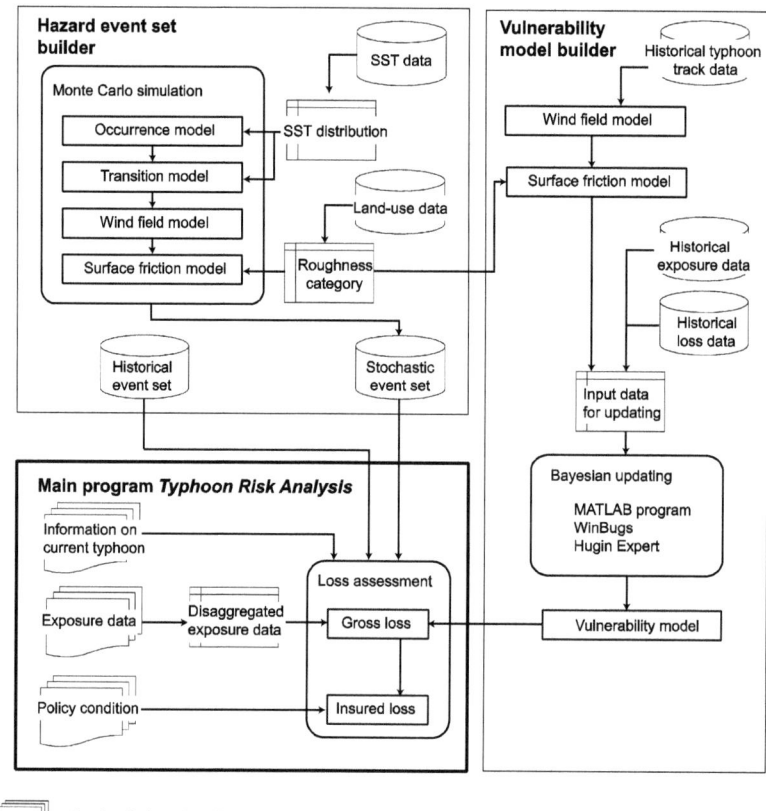

Figure 6.9: Components of programs.

6.4. *TRAST*: Software tool for portfolio risk

This section explains the functions of the developed software tool *TRAST* and shows the analysis results which can be produced by *TRAST*. Figure 6.10 shows the main screen of *TRAST*. To perform a analysis the user has to do the following four steps. First the portfolio which has to be analyzed has to be imported as exposure data, then the policy conditions have to be defined, then the portfolio has to be disaggregated as described in Section 6.3.3 and finally the conditions for the analysis have to be defined before the analysis can be performed.

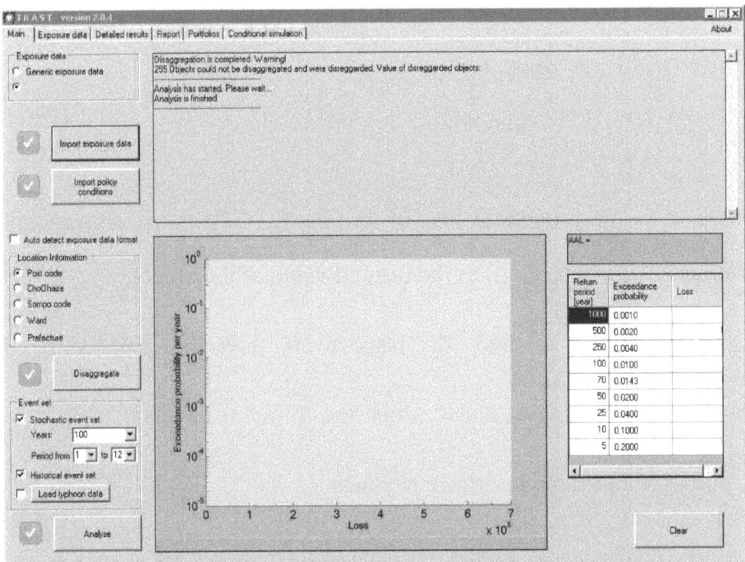

Figure 6.10: Graphical user interface (*Main* tab window).

6.4.1. Defining the policy conditions

The software tool *TRAST* supports two different formats for defining the policy conditions as described in the following. The definition of the policy conditions are parameterized so that *TRAST* also can be used for a sensitivity analysis for the parameters and to optimize the parameters describing the policy conditions.

Format "A"

The format "A" can be utilized to describe the policy conditions characterized by the following functional form (see also Figure 6.11):

$$f(x) = \begin{cases} 0 & (x < \min(V_{franchise}, \alpha_1 V_{insured})) \\ k_1 x & (\min(V_{franchise}, \alpha_1 V_{insured}) \leq x < \alpha_2 V_{insured}) \\ k_2 x & (\alpha_2 V_{insured} < x) \end{cases} \quad (5.31)$$

where x is the ground-up loss of an object, $f(x)$ is the corresponding insurance payment, $V_{insured}$ is the insured value of the object, $V_{franchise}$ is the franchise value, α_1 and α_2 respectively are the parameters that characterize the criteria under which the different payment rates k_1 and k_2 are applied for the insurance payment.

6. Application: Portfolio risk analysis

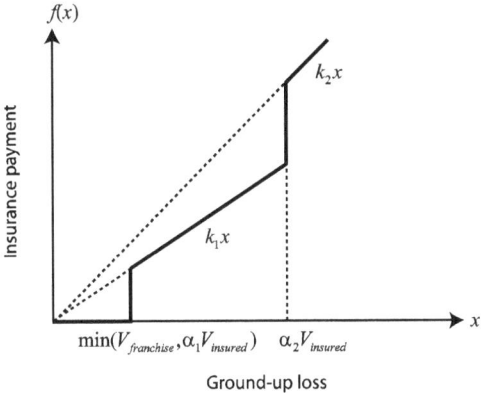

Figure 6.11: Insurance payment of the policy conditions described by the policy data file in the format "A".

Format "B"

The format "B" can be utilized to describe the policy conditions characterized by the following two functional forms (see also Figure 6.12):

$$f_0(x) = \begin{cases} 0 & (x < V_{franchise}) \\ k_1 x & (V_{franchise} \leq x < \alpha_2 V_{insured}) \\ k_2 x & (\alpha_2 V_{insured} \leq x) \end{cases} \quad (5.32)$$

and

$$f_1(x) = \begin{cases} 0 & (x < \alpha_1 V_{insured}) \\ k_1 V_{insured} & (\alpha_1 V_{insured} \leq x < \alpha_2 V_{insured}) \\ k_2 V_{insured} & (\alpha_2 V_{insured} \leq x) \end{cases} \quad (5.33)$$

where x is the ground-up loss of an object, $f_0(x)$ and $f_1(x)$ are the insurance payments corresponding to two types of policies that are

-173-

explained below, $V_{insured}$ is the insured value of the object and $V_{franchise}$ is the franchise value, which appears only in $f_0(x)$. α_1 (only in $f_1(x)$), α_2, k_1 and k_2 are the parameters that characterize the policies.

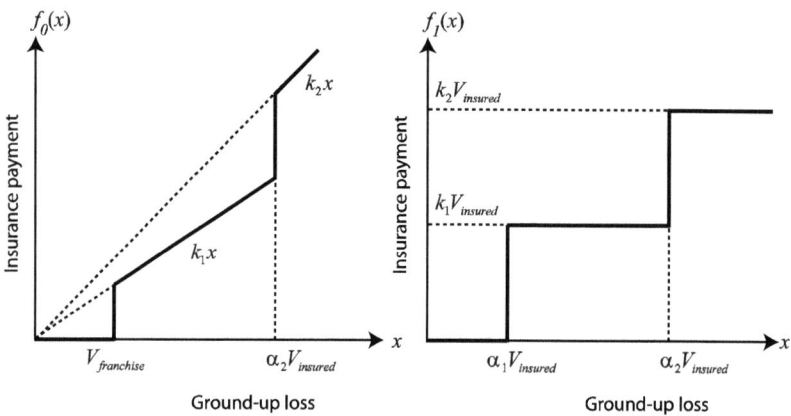

Figure 6.12: Insurance payments of the two policy conditions described by the policy data file in the format "B".

6.4.2. Defining the analysis conditions

The conditions for the risk analysis can be defined in the user interface of *TRAST*. The statistics of the annual insured loss, can be evaluated by selecting the "Stochastic event set" (①in Figure 6.13), whereby also the number of years of the simulation for the analysis has to be specified. It is recommended to enter a number more than 10000 for the stable evaluation of the statistics.

6. Application: Portfolio risk analysis

To evaluate the statistics of the insured losses of a portfolio due to the historical typhoon events that are stored in the typhoon event sets[22], the "Historical event set" (② in Figure 6.13) can be used.

In the case when the statistics of the insured loss of a typhoon event that is not stored in the typhoon event set should be evaluated, a typhoon track data file can be imported in which the typhoon event is defined (③ in Figure 6.13).

The analysis will be performed as soon the button "Analyze" (④ in Figure 6.13)[23] is clicked.

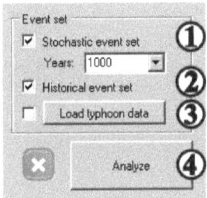

Figure 6.13: Components concerning the analysis settings.

---

22 Typhoons that made landfalls to Japan between 1971 and 2004 and the typhoon Vera are included in the historical event set.
23 The analysis time required for the analysis depends on the number of years of the simulation as wells as the size of the portfolio; for example, the analysis of the portfolio consisting 15 million objects with the 10000-year simulation on a computer with an Intel core2duo 2.4 GHz approximately takes 48 hours.

6.4.3. Displaying analysis results

The summary of the analysis results is displayed in the tab window "Main", if the analysis is performed with the stochastic event. The summary
includes the following statistics (see Figure 6.14):

- Exceedance probability (EP) curve of the insured loss (①)
- Annual average loss (AAL) (②)
- The expected loss for the return periods of 5, 10, 25, 50, 70, 100, 250, 500 and 1000 years (③)

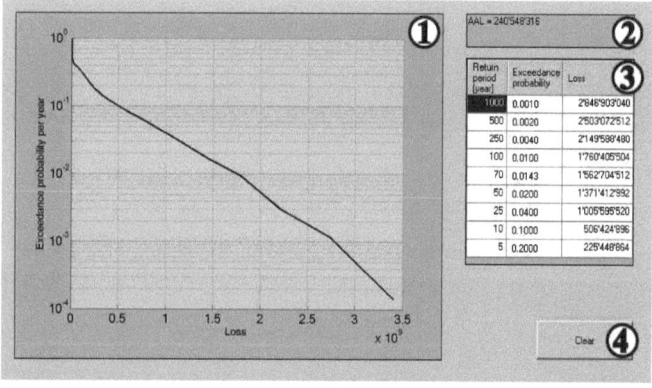

Figure 6.14: Summary of the analysis results for the case of the analysis with the stochastic event set.

Detailed results

The detailed results of the analysis are displayed in the tab window "Detailed result" (See Figure 6.15). The detailed results are available for

6. Application: Portfolio risk analysis

all types of the analysis setting. In this tab window, the following types of analysis results are available:

- Statistics of the insured loss of the individual typhoon events in the table "Stochastic event set", if the analysis is performed with the stochastic event set, or with the options "conditional simulation" (the conditional simulation is explained in Section 8.1).
- Statistics of the insured loss of the individual typhoon events in the table "Historical event set", if the analysis is performed with the historical event set, or with the options "scenario-based simulation".
- Graphical representation of the typhoon track and maximum wind speeds of the individual typhoon events as well as graphical distribution of the exposures of the portfolio.

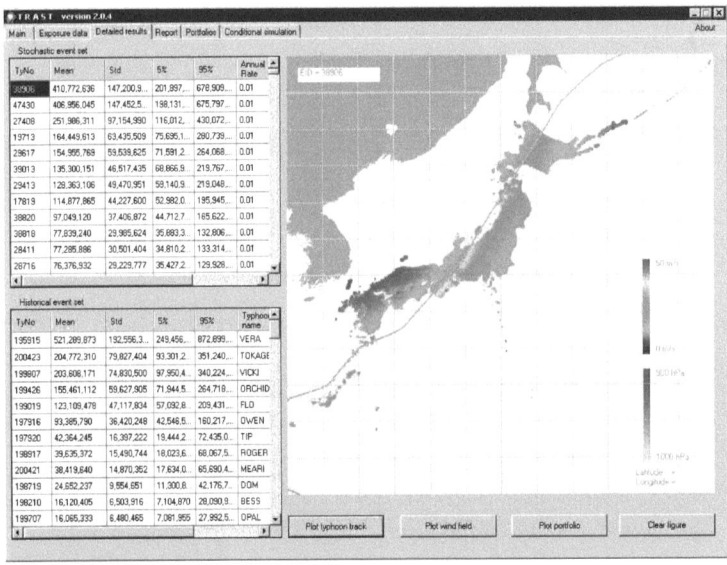

Figure 6.15: Losses of individual typhoon events and visualization of track and wind speeds of a selected event.

6. Application: Portfolio risk analysis

The explanation of each part in this tab window is given in the following sections.

Analysis results in the table "Stochastic event set"

This table contains the analysis results when the analysis is performed with the stochastic event set or with the options "scenario-based simulation". Each row of the table contains the statistics of the insured loss due to each typhoon event:

- TyNo: identification number of the typhoon event
- Mean: mean value of the insured loss due to the typhoon event
- Std: standard deviation of the insured loss due to the typhoon event
- 5%: 5%-quantile of the insured loss due to the typhoon event
- 95%: 95%-quantile of the insured loss due to the typhoon event
- Annual rate: annual occurrence rate of the typhoon event

Analysis results in the table "Historical event set"

This table contains the analysis results when the analysis is performed with the historical event set. Each row of the table contains the statistics of insured loss due to each typhoon event in the historical event set:

- TyNo: identification number of the typhoon event
- Mean: mean value of the insured loss due to the typhoon event
- Std: standard deviation of the insured loss due to the typhoon event
- 5%: 5%-quantile of the insured loss due to the typhoon event
- 95%: 95%-quantile of the insured loss due to the typhoon event
- Typhoon name: name of the typhoon event

Visualizing the typhoon events

The typhoon track of a typhoon event (in the stochastic and historical event sets) can be visualized. Note that the line color of the typhoon track corresponds to the value of the central pressure of the typhoon at each respective time. The maximum wind speeds at locations on the Japanese islands of a typhoon event can be visualized on the map The geographical distribution of the portfolio for which the insured loss is analyzed can be visualized. The color of the dots corresponds to the value of the sum of the insured values of the exposures in the locations. The typhoon track, maximum wind speeds and geographical distribution of the portfolio can be superposed.

In the Appendix D 10.4 are several examples of the visualization of the typhoon track and the corresponding wind field shown.

6. Application: Portfolio risk analysis

Generating a report

The software tool *TRAST* includes a automatic report generator, which summarizes the analysis results. The report on the summary of the analysis can be saved in a selected format (Acrobat pdf, Microsoft Word and Excel) (see Figure 6.16).

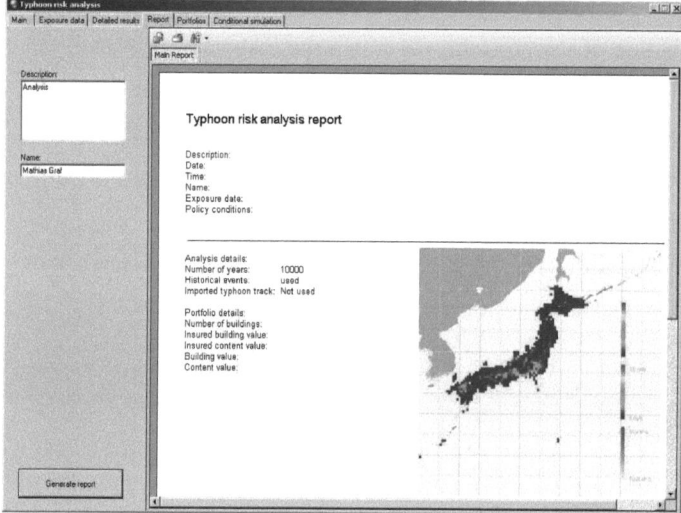

Figure 6.16: Report generator.

Managing analysis results

The software tool *TRAST* allows also to manage the analysis (Figure 6.17). Two main functions are provided for the management of the analysis results:

- Save, load and delete the analysis results
- Combine the analysis results

Combining the analysis results refers that the analysis results of two or more different individual portfolios are combined so that the statistics of the insured loss of the integrated portfolio that consists of the individual portfolios are evaluated. It is also possible to combine the results of the analyses that are performed on several PCs.

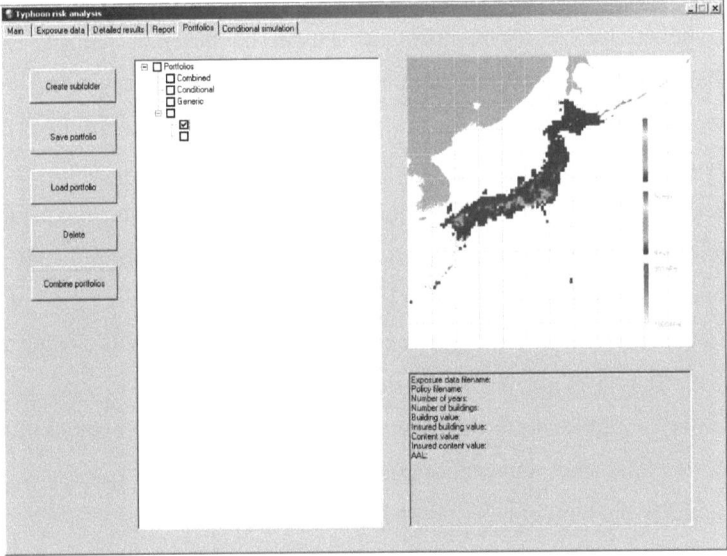

Figure 6.17: Interface for the management of the analysis results.

7. Application: Global warming risk assessment

Another potential application of the developed model is the investigation of the effects of global climate change on the probabilistic characteristics of strong wind speed induced by typhoons.

Emanuel (1987) compares the mechanism behind the intensity of a typhoon to a Carnot cycle. The intensity of a typhoon is dependent on the difference between the SST and the temperature in the troposphere. The typhoon absorbs energy from the warm SST and cools down in the higher atmosphere. The intensity of a typhoon gets higher if the temperature difference is higher. Depending on how climate change will influence this balance, the typhoon activity can increase or decrease. Observation of the historical data shows that typhoons occur only if the SST is higher than 26.5°C. It is a major debate in the research community whether global warming would increase the number of typhoons or not. Emanuel (2005) proposes that the number of strong typhoons will increase but not necessarily the total number but the number of typhoons.

A preliminary study on this is undertaken in (Graf et al., 2008). This study describes a sensitivity analysis of the influence of the SST in the probabilistic typhoon model. Assuming that the developed probabilistic typhoon model can be used to predict the typhoon activity in a future climate scenario and that the climate change is only represented by the SST, this study investigates the impact of the change of the SST on the maximum wind speeds and wind loads and presents an approach on how the effect of the climate change can be investigated. In this study, assuming several scenarios for the future change of SST, it is found that the upper quantile values of the annual maximum wind speed distribution may significantly increase as a consequence of an increase in the SST; in turn, the probability of failure of structures due to wind loads may also significantly increase. This study is presented in Section 7.1.

The second study, described in Section 7.2, presents a more realistic investigation of the effect of global warming by considering the increase in the SST and the temperature of the atmosphere for future climate scenarios by using the output of the mesoscale meteorological model JMA-NHM. In cooperation with Professor Takashi Maruyama from Kyoto University, the data obtained from the mesoscale meteorological model JMA-NHM is used to establish the typhoon model using the model builder described in Section 5.6. This in turn is used to estimate the change in the wind risk of residential buildings in Japan under a future climate (Nishijima et al., 2011). This application shows how the Bayesian framework can be used to assess the effects of global warming by updating the typhoon model with new available data.

7.1. Adaption of typhoon risk modelling to climate changes

Latest analyses of the effect of global warming indicate that increased sea surface temperatures (SST) will increase the frequency of tropical cyclones with extreme wind speeds; for a general overview see the Intergovernmental Panel on Climate Change report (Solomon et al., 2007) or the report of (World Meteorological Organization, 2006). Knutson and Tuleya investigate the effect of global warming on tropical cyclone activity on the basis of a selection of different global climate models and tropical cyclone models, assuming different scenarios for CO_2 emissions (Knutson and Tuleya, 2004; Knutson et al., 2010). The analyses suggest that over a time frame of 80 years on average the minimum central pressure of tropical cyclones will decrease by about 14% from its present value and the maximum wind speed will increase by about 6%. The results from such analyses surely depend on the utilized models concerning climatic changes, tropical cyclones as well as the considered scenarios concerning emission of greenhouse gases and are thus subject to significant uncertainties. However, it is fair to assume that future tropical cyclones will result in more frequent strong wind events (Emanuel, 1987; Emanuel, 2005). This in turn will increase wind loads on buildings and structures in general and induce an increase of damages to the built environment.

7.1.1. Objectives of the study

Whereas the effects of global warming are likely to induce an increase of damages to the built environment, the magnitude of such consequences and the effectiveness of implementing new policies on the design and maintenance of the built environment are not yet obvious. Motivated by this and focusing on structures as an example of the constituents of built environment the study addresses the following two questions; 1) to what degree will the reliability of structures be decreased due to the intensified tropical cyclone activity? and 2) to which extent should the policies regarding design and maintenance of structures be changed in order to maintain the present level of reliability? Answers to these questions are investigated in the present study for structures located in the region of the northwest Pacific. However, the approach adopted herein can be applied to other regions as long as appropriate data and models are available.

7.1.2. Approach adopted in the study

For the purpose to assess the reliability of structures subject to wind loads the study takes basis in a probability-based engineering approach, see e.g. (JCSS, 2001). The approach requires a probabilistic hazard model and a probabilistic fragility model; each respective model is utilized to calculate the probability distribution of wind loads and the probability distribution of the resistance of structures subject to wind loads. The hazard model adopted in the present study is the probabilistic typhoon model developed during this thesis, see (Graf et al., 2008), which enables to incorporate the effect of different SSTs on the occurrence and development of typhoons. The fragility model postulated in the present study is based on the JCSS Probabilistic Model Code, see (JCSS, 2001) The roles of these models are shown in Figure 7.1 in the context of the assessment of structural reliability. The figure shows the analysis flow of the adopted approach to; a) assess the effect of global warming on the reliability of structures and b) identify the necessary change of design and assessment policies for structures, required to maintain the present level of structural reliability.

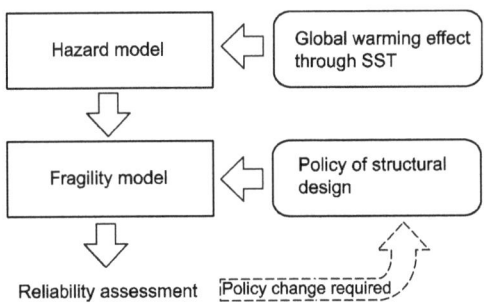

Figure 7.1: Approach suggested to assess and mitigate the impact of extreme wind events caused by climatic change.

7.1.3. Validation of the adopted hazard model for wind loads assessment

The probabilistic typhoon model utilized in the assessment of structural reliability enables to evaluate wind loads at different locations in probabilistic terms; e.g. the probability distribution of annual maximum wind loads. However, since the probabilistic model is established on the basis of not only scientific knowledge but also statistical analyses using available historical data it is not sure to which extent the model may be extrapolated beyond the historically observed SSTs. Therefore, in order to assess the plausibility of the model and to apply this in further assessments simulation results from the probabilistic model must be validated with other analysis results. In the present study the simulation results are compared with the analysis results which do not depend on the historical data, e.g. (Knutson and Tuleya, 2004).

7.1.4. Assessment of the reliability of structures

Applying the probabilistic typhoon model for each scenario of the SST increase, the probability distribution of annual maximum wind load s for a structure is obtained. The distribution is represented by the probability density function $f_S(s \mid e_{SST})$ conditional on a given scenario e_{SST}. The probability distribution of the resistance R of a structure is modelled in accordance with the JCSS Probabilistic Model Code in terms of the conditional probability density function $f_R(r \mid \mu_R)$. μ_R represents the expected value of the resistance of the structure and is related to the policy regarding the design and maintenance of the structure. Thus the structural performance is a function of the policy. The annual probability of failure

7. Application: Global warming risk assessment

$P_F(e_{SST}, d)$ and the so-called reliability index $\beta(e_{SST}, d)$ for the structure are calculated as a function of the scenario and the policy as:

$$P_F(e_{SST}, \mu_R) = \int_{\{r-s<0\}} f_S(s \mid e_{SST}) f_R(r \mid \mu_R) ds dr, \qquad (6.1)$$

$$\beta(e_{SST}, \mu_R) = -\Phi^{-1}\left(P_F(e_{SST}, \mu_R)\right), \qquad (6.2)$$

where $\Phi^{-1}(\cdot)$ is the inverse of the standard normal cumulative distribution function. The reliability index is widely used in the structural engineering field and the appropriate levels of the reliability index are suggested in the JCSS Probabilistic Model Code as a function of the significance of the structure and the relative cost for installing safety measures.

Finally, the decrease of the reliability of a structure is calculated from Equation (6.2). Furthermore, the policy in regard to the design and assessment of a structure which is necessary in order to maintain the present level of reliability is identified in terms of the expected value of the resistance μ_R^* for a given scenario e_{SST}^* through the following equation:

$$\beta(e_{SST}^*, \mu_R^*) = \beta(e_{SST}^0, \mu_R^0) \qquad (6.3)$$

where e_{SST}^0 represents the scenario that the SST remains at the present level, and μ_R^0 represents the expected value of the resistance corresponding to the present level of reliability.

7.1.5. Example

In the following, the proposed approach is illustrated in a example. This example investigates the effect of global warming represented by a increase in the sea surface temperature (SST) on the wind speeds induced by typhoons and the effect on the probability of failure of buildings. For this study it is assumed, that the frequency and the location of the occurrence of typhoons do not change due to global warming. Several typhoon simulations are performed whereby in each simulation the available STT map of the year 2000, which is a input parameter to the transition model (see Section 2.6), is increased step wise.

To assess the change of the probability of failure due to the increased wind speeds, a optimally designed building is assumed, which has a target annual probability of failure of 10^{-5} as proposed in the JCSS Probabilistic model code (JCSS, 2002). The resistance of the building is represented with a log normal distribution as $LN(\lambda, \zeta)$ (Melchers, 2001). The parameter ζ is assumed to be 0.2 and the parameter λ is optimized so that in combination with the wind load obtained from the typhoon simulations the building has a probability of failure of 10^{-5} (see Figure 7.2). The wind load is calculated by $L = kV^2$, whereby V is the annual

7. Application: Global warming risk assessment

maximum wind speed and k is a factor which is assumed to be 1 for this example.

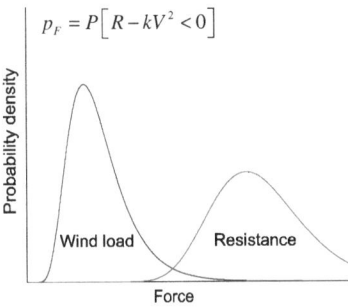

Figure 7.2: Estimation of the target probability of failure.

The optimally designed building in this example is located in Osaka and the typhoon model described in Chapter 2 is used in combination with the increased SST map to assess the wind speeds and the probability of failure for this building. A higher SST leads to a lower central pressure of the simulated typhoon, which in turn leads to a higher wind speed. The increase in the 98% quantile of the annual wind speed is shown in Figure 7.3 (left) and the resulting change of the probability of failure is shown in Figure 7.3 (right). According to this study an increase of the SST by 2° increases the probability of failure by factor of ten.

7. Application: Global warming risk assessment

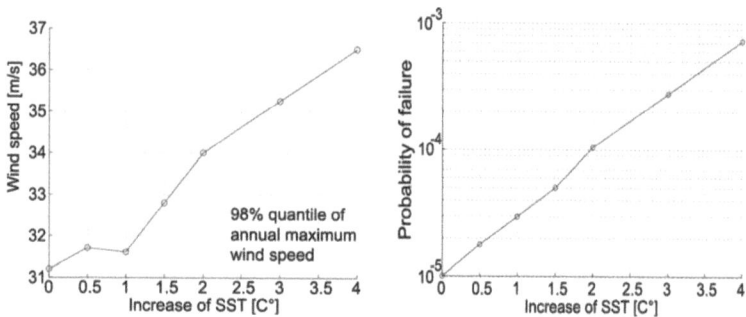

Figure 7.3: Change of the characteristic value (98%-quantile value) of annual maximum wind speed (left) and the change of the probability of failure (right).

A change of the design policy may be required to maintain the target reliability as shown in Figure 7.4 (left). Figure 7.4 (right) shows how much the 5% quantile of the resistance of the building has to be increased to maintain the target probability of failure of 10^{-5}.

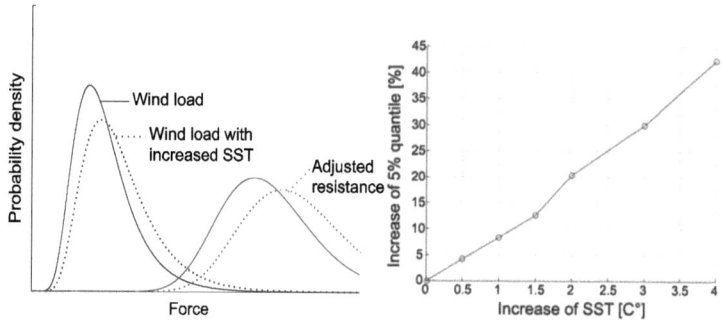

Figure 7.4: Adaption of structural design (left) and required change of the characteristic value (right).

In (Burri et al., 2009) a study was conducted which uses the output of different climate projection models described in the IPCC Report of 2007 (Solomon et al., 2007) as SST maps and considers the whole Japanese Islands.

7.1.6. Summary

The present study investigates the effect of global warming on structural reliability in the context of a possible increase of tropical cyclone activity. For this purpose a probability-based engineering approach is adopted. The approach employs the probabilistic typhoon model that is developed during this thesis. The probabilistic model for the resistance of structures is adapted from the JCSS Probabilistic Model Code. First the consistency of the probabilistic typhoon model is verified with the results of alternative models not relying on historical data. Thereafter the suggested model is applied for assessing the change of structural reliability considering the effect of the increased SST on tropical cyclone activity. Thereby it is also investigated to which extent the resistance of structures must be increased in order to maintain the present level of structural reliability. Although these investigations are made for structures in the northwest Pacific region the approach adopted in the present study can be applied to other regions to the extent that relevant models and data are available.

7.2. A preliminary impact assessment of typhoon wind risk of residential buildings in Japan under future climate change

This section investigates a quantitative impact assessment of the climate change on typhoon wind risk, focusing on residential buildings in Japan. The risk is assessed based on (1) the typhoon event set extracted from the simulation by the super-high resolution atmospheric general circulation model developed within the KAKUSHIN program; (2) the probabilistic typhoon modeling scheme developed during this thesis; (3) a damage ratio model empirically estimated on the basis of the damage report of typhoon Songda (2004) and the reproduced wind field by a mesoscale meteorological model; JMA-NHM. The main results are that in the future (2075-2099) at most locations of Japan: (1) extreme wind events (10-minutes sustained wind speed >30m/s) are more likely to occur; (2) the median of the annual maximum wind speed decreases; (3) the expected number of damaged residential buildings decreases, assuming that the profile of the building portfolio remains unchanged. Based on these results, the assumptions and inputs employed in the assessment are critically reviewed. Thereby, the needs of further research efforts toward more credible and comprehensive impact assessment are addressed.

7.2.1. Introduction

Global-scale meteorological monitoring networks have revealed that the global climate has changed significantly over the last decades. Large amount of scientific work has claimed that further change of the climate may occur in the future. In response to these, various political as well as non-political protocols on action plans for mitigating the climate change have been proposed at different levels of the society; only to find it difficult to reach the envisaged goals. Presently, there seems a general agreement that actions have to be undertaken also on the adaptation of the society to the emerging global climate change.

For the purpose to facilitate decisions on the adaptive actions, a large amount of effort have been and is being devoted worldwide to develop more credible models for the projection of the future climate change - numerical models of the climate as well as models for the growth of the human society. Correspondingly, quantitative assessment of the impact of the climate change on the human society has become possible for various types of risks; among others, the risks due to natural hazards affected by the climate change.

This section investigates the probabilistic assessment of the impact of the climate change on infrastructure, using the super high-resolution general circulation model developed within the KAKUSHIN program. Focus is given on the typhoon wind risk of residential buildings in Japan. The objectives are: (1) to demonstrate that tools to assess the impact are readily available, if not yet these are perfect; thereby, (2) to address the issues and needs for further sophistication of the tools for the purpose to facilitate more credible and comprehensive assessment of the risks.

7.2.2. Approach

Procedure for the impact assessment

The impact of the climate change in the future on the typhoon risk of residential buildings in Japan is assessed in the following procedure. First, typhoons simulated with the atmospheric general circulation model (AGCM) under current and future climates are extracted. Second, based on the extracted typhoons, probabilistic typhoon hazard models are developed for each of the climates. Third, using the developed probabilistic typhoon hazard models, stochastic typhoon events are simulated by Monte Carlo simulations. Finally, the risks are assessed for both climates, and the rate of the change of the risk is calculated. A simplistic measure of risk is assumed in the present study, which is defined as the damage ratio of the residential buildings. In the following, each of the step is explained.

Typhoons extracted from AGCM simulation

The present study utilizes the simulation results with the AGCM, which has been developed under the project "Projection of the change in future weather extremes using super-high-resolution atmospheric models" (Mizuta et al., 2011) within the framework of KAKUSHIN program. The resolution of the AGCM is 20-km mesh grid. The simulations are performed for three different periods; i.e. 1979-2003 (*current*), 2015-2039 (*near future*), and 2075-2099 (*future*). For the simulations, the dataset corresponding to the IPCC AR4 A1B scenario is employed as the boundary conditions to the AGCM. The abovementioned project consists of two phases, and correspondingly two sets of the simulation results are available, which are labeled as MRI-AGCM3.1S and MRI-AGCM3.2S respectively. In the present study, the latter set is utilized, and the impact

7. Application: Global warming risk assessment

assessment is undertaken for the *future* climate relative to the *current* climate.

Typhoons are extracted from the simulation results, see (Murakami and Sugi, 2010). The extracted typhoons are represented in terms of the central pressure as well as the position of the typhoons at each time. Figure 7.5 illustrates the extracted typhoons. These are utilized as the basis for the development of the probabilistic models for the occurrence and the transition of typhoons.

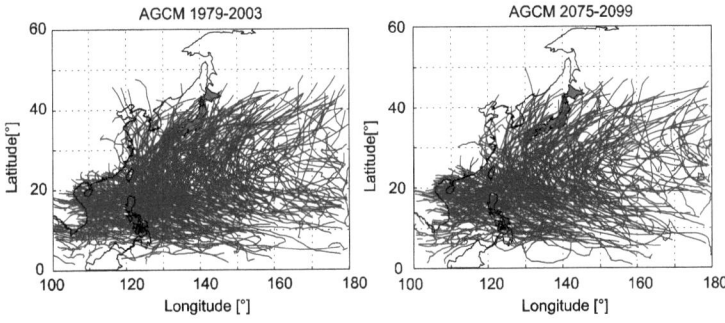

Figure 7.5: Extracted typhoons for the current climate (left) and the future climate (right).

Probabilistic modeling of typhoon

The present study takes basis in the probabilistic typhoon hazard modeling proposed by (Graf et al., 2009) and (Graf and Nishijima, 2011). The model consists of four component models; i.e. occurrence model, transition model, wind field model and surface friction model. The first two models are probabilistic models, whereas the last two are deterministic ones. In the present study, two versions of the occurrence and transition models are developed for the current and future climates using the corresponding

AGCM simulation results; on the other hand, the models developed based on the methodology presented in (Graf et al., 2009) are employed without modifications for wind field and surface friction modeling. The outputs from the suit of the hazard models are the probabilistic characteristics of the maximum 10-minute sustained wind speeds at each location in Japan during individual typhoon events and their frequencies.

It should be emphasized here that the occurrence and transition models for the current climate are developed based on the typhoons subtracted from the AGCM simulation, instead of the actual historical typhoons archived in e.g. the JMA Best track data. Thereby, it is anticipated that the effects of the biases inherited in the AGCM on the assessments are alleviated.

Modeling of performance of residential buildings

The assessments of the typhoon wind risk require a model that represents the performance of structures as a function of wind speed. The typhoon Songda in 2004 is selected to develop such a model, since a report on the damage on the residential buildings as well as the observation of wind speed at NeWMeK stations are available. The observation of wind speed is utilized for the calibration of wind speed.

The statistics are reported on damaged residential buildings at municipality level in Kyushu, Japan, caused by the typhoon Songda in 2004, see (Tomokiyo et al., 2009). The report on the damage differentiates the degree of damages, i.e. minor, moderate and major damages, and summarizes the number of damaged buildings for each class of damages at each municipality. These statistics are utilized, together with the Census data, as the basis for developing the model; hereafter, the model is called damage ratio model. Note that the damage ratio model developed here

7. Application: Global warming risk assessment

does not differentiate the degree of damages. The damage ratio model describes the ratio of the number of damaged buildings over the total number of households (which approximates the number of residential buildings) in a given area as a function of the maximum wind speed during a typhoon event.

In order to develop the damage ratio model, the wind speeds at respective municipalities are required. The strong wind field of the typhoon Songda is reproduced numerically using the JMA-NHM (Japan Meteorological Agency Non-Hydrostatic Model) together with the JMA-RANAL data as the initial condition and the JMA-RSM data as the boundary condition, see (Maruyama et al., 2010). In Figure 7.6, the dots represent the pairs of the actual damage ratio and the maximum wind reproduced by the JMA-NHM at each location. The solid line represents the estimated damage ratio model. The damage ratio model is developed by a regression analysis, and is expressed as:

$$r_d(y_{NHM,MAX}) = 5.1 \cdot 10^{-9} \cdot y_{NHM,MAX}^{4.0} \tag{6.4}$$

where $r_d(y_{NHM,MAX})$ is the damage ratio as a function of $y_{NHM,MAX}$, the maximum wind speed at 10 [m] height from the surface reproduced by the JMA-NHM. It is assumed in the present study that the model developed in this way is representative for other typhoon events as well as other regions in Japan.

7. Application: Global warming risk assessment

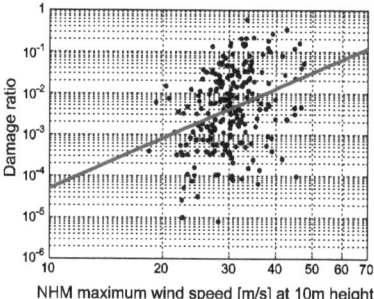

Figure 7.6: Relation between damage ratio and the maximum wind speed $y_{NHM,MAX}$ computed with the JMA-NHM at different locations and the estimated damage ratio model.

Calibration of wind speeds

The wind speed computed with the probabilistic hazard model corresponds to 10-minute sustained wind speed, whereas the wind speed computed with the JMA-NHM is considered as wind speed spatially and temporally averaged; the extents of which, however, are not explicitly known. For this reason, a calibration of the wind speeds between the JMA-NHM computation and the NeWMeK observation is carried out by (Maruyama et al., 2010). The relation for the calibration is expressed as:

$$y_{NewMeK,10\min,MAX} = 0.83 \cdot y_{NHM,MAX} \quad (6.5)$$

where $y_{NewMeK,10\min,MAX}$ is the maximum observed 10-minute sustained wind speed at a NeWMeK observation station. Note that the wind speed computed by the probabilistic models introduced in the previous section is calibrated to the maximum 10-minute sustained wind speed, using the observations at JMA meteorological stations. Assuming that the observational characteristics of the NewMeK and JMA stations are identical, the probabilistic hazard model and the damage ratio model can be then integrated.

Measure of risk

In this section, the measure of risk assumed in the present study is explained. Let N_j^B and N_j respectively denote the number of buildings at location j and the number of buildings that are damaged by typhoons in a given year. The number N_j is a random variable and can be written as:

$$N_j = \sum_{k=1}^{K_j} N_{j,k} \qquad (6.6)$$

where $N_{j,k}$ is the number of damaged buildings in the event of the kth typhoon in a year, and K_j is the number of typhoon events in the year relevant to the buildings at location j, which is also a random variable. The definition of the building risk $\bar{R}_j$ at location j is given as the expected value of the number N_j relative to the total number N_j^B of the buildings:

$$\bar{R}_j = \frac{E[N_j]}{N_j^B} = E\left[\frac{N_j}{N_j^B}\right] = E\left[\sum_{k=1}^{K_j} \frac{N_{j,k}}{N_j^B}\right] = E\left[\sum_{k=1}^{K_j} R_{j,k}\right] = E[K_j] \cdot E[R_{j,k}] \qquad (6.7)$$

where, $R_{j,k} = N_{j,k} / N_j^B$. Here, it is assumed that $R_{j,k}$ are independently identically distributed. $E[R_{j,k}]$ is defined using the probability density function $f_{Y_j}(y)$ of the maximum wind speed Y_j at location j in any given single typhoon event, and the building damage ratio $r_d(y)$ as a function of the wind speed y as:

$$E\left[R_{j,k}\right] = \int r_d(y) f_{Y_j}(y) dy \tag{6.8}$$

Thus, given the building damage ratio $r_d(y)$, the building risk $\bar{R}_j$ at location j is characterized by two wind hazard components: the mean frequency $E[K_j]$ of typhoon events relevant to location j in a given year, and the probability density function $f_{Y_j}(y)$ of the maximum wind speed in an event.

7.2.3. Results

Simulation of typhoon events and its verification

Using the two sets of the probabilistic hazard models, two corresponding sets of the typhoon events are generated by Monte Carlo simulations, for the current climate and the future climate. Each set contains typhoon events corresponding to 25000 times of one-year simulation. For each typhoon event, the maximum 10-minute sustained wind speeds are computed at municipality level entirely covering Japan; i.e. at 2249 locations.

The verification of the generated typhoon events is undertaken by comparing a couple of statistics of the transition of the typhoons. In Figure 7.7, the cumulative annual average numbers of typhoons that travel through as a function of the central pressure at latitudes of 30N° and 35N° under the current and future climates are shown. It can be said that the overall performance of the probabilistic models is acceptable in the sense that these succeed in representing two general tendencies observed from the simulations with the AGCM: (1) the overall number of typhoons travelling though these latitudes is decreasing; (2) the number of intensified typhoons is increasing. However, there is a discrepancy in the numbers of weaker typhoons for the latitude of 35N° under the future climate. The main reason for the discrepancy is considered due to the smaller numbers of typhoons in the simulation with the AGCM to develop the probabilistic model.

7. Application: Global warming risk assessment

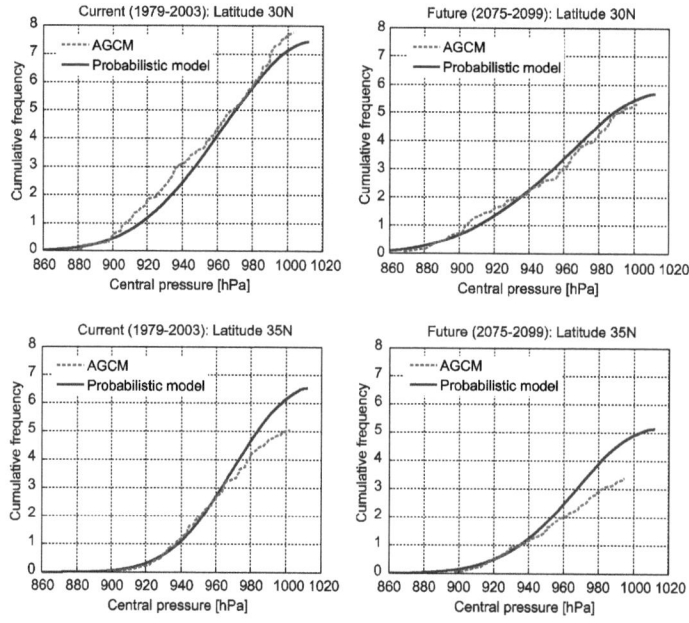

Figure 7.7: Cumulative annual average numbers of typhoons in the AGCM simulation and the Monte Carlo simulation as a function of the central pressure at latitudes 30N° (top) and 35N° (bottom) under the current climate (left) and the projected future climate (right).

Impact on annual maximum wind speed

Having verified the performance of the probabilistic models relative to the simulations with the AGCM, the impact of the climate change on the annual maximum wind speed is investigated. In Figure 7.8, the changes of the distributions of the annual maximum wind speeds at two locations in Japan (Tokyo and Fukuoka) are shown. Note that these wind speeds are the ones at 10 m height from the surface computed considering actual roughness conditions; the roughness conditions are considered using land use data in terms of roughness length; see (Graf et al., 2009) for the way how the roughness length is empirically estimated using land use data.

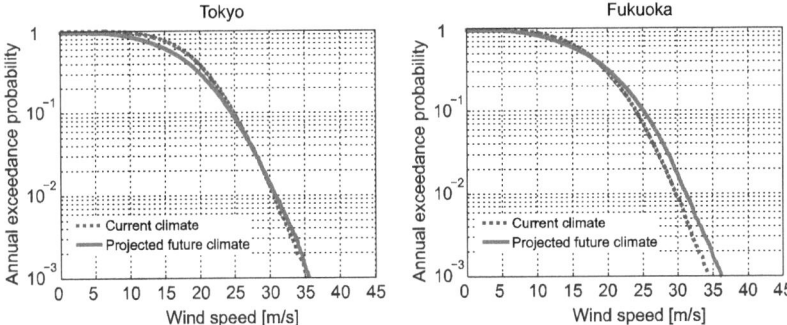

Figure 7.8: Exceedance probabilities of the annual maximum wind speeds under the current and the projected future climates at Tokyo (left) and Fukuoka (right).

It is seen that the probabilities of the occurrence of higher annual maximum wind speeds (>30m/s) increase at all three locations, whereas the medians of the annual maximum wind speeds decrease. This is consistent with the two observations from the simulations with the AGCM mentioned in the previous section.

Furthermore, the overall comparisons of the 98%-quantiles (corresponding to 50-year return period) and the medians of the annual maximum wind speeds at 2249 locations in Japan are made, see Figure 7.9, confirming that these observations are true for most of the locations in Japan.

7. Application: Global warming risk assessment

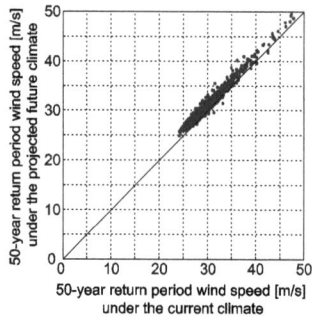 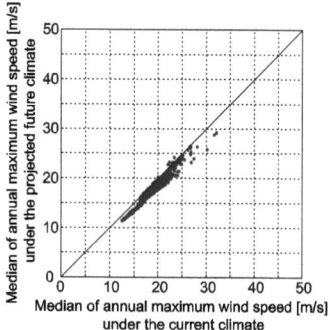

Figure 7.9: Comparisons of 50-year return period wind speeds (left) and medians of the annual maximum wind speeds (right) at 2249 locations in Japan under the current and the projected future climates.

Impact on wind risk of residential buildings

The risks of residential buildings are computed at each location in Japan under the current and the future climates. The result is presented in Figure 7.10. As seen in the figure, the risks tend to decrease for most of the locations. The main reason for this is the decrease in the frequency of the relevant typhoon events in Japan and the degree of the intensification of typhoons is not enough to compensate this degrease. The simple average of the change rates of the risks is 0.87. The geographical distribution of the change is shown in Figure 7.11, which indicates the tendency that the risks increase in the northwestern part of Japan and decrease in the southern eastern part. This is due to the change of the characteristics of the track of typhoons.

7. Application: Global warming risk assessment

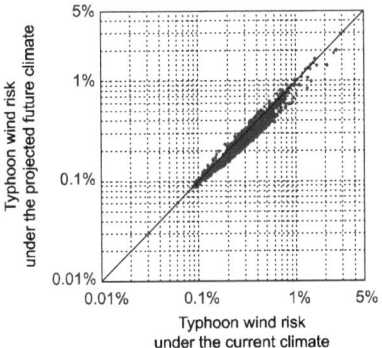

Figure 7.10: Change of the typhoon wind risks under the current and projected future climates.

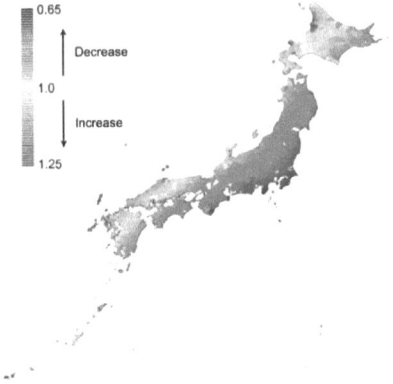

Figure 7.11: Geographical distribution of the change of the residential building risks.

7.2.4. Discussion

General issues

The numerical results obtained in the present study depend on several critical inputs and assumptions. Among others, the probabilistic characteristics of typhoon events are crucially dependent on the outcomes from the AGCM and the employed scenario as well as the damage ratio model. Furthermore, the numbers of typhoons in these outcomes may not be sufficient to develop credible probabilistic hazard models; the assumption that the structural performance of building is uniform over Japan may not be appropriate, since among others the building code differentiates the wind design loads for buildings in different regions of Japan. Improvements on these are addressed as future tasks.

It should be remarked that the impact assessment of the risk is subject to the definition of risk. For instance, defining the measure of risk in monetary terms, instead of the damage ratio as assumed in the present study, might result in a different conclusion, since it is anticipated that the monetary loss may be highly nonlinear to the wind speed. However, in order to facilitate the quantitative assessment of this, a vulnerability model that describes the relation between wind speed and monetary loss needs to be developed.

In the following sections, a few specific issues are discussed in order to identify clear direction of the research needs toward more credible and comprehensive assessment.

7. Application: Global warming risk assessment

AGCM Performance in regard to the simulation of typhoons

The performance of the AGCM in regard to the simulation of typhoons is investigated by comparing some statistics on the typhoons in the AGCM simulations and in the JMA Best track data in the same period; i.e. 1979-2003. The statistics employed for this purpose are the annual average numbers of typhoons that travel through the several latitudes with certain values of the central pressure or smaller. Figure 7.12 shows the comparison results for the latitudes of 20N° and 35N°. At 20N°, the statistics are in good agreement with each other; at 35N °, significant discrepancy is observed. It should be mentioned, however, that the impact assessment in the present study is performed based on the simulations with the AGCM for the current climate and also for the future climate; hence, the biases associated with the AGCM are expected to be alleviated to some extent. Clearly, further improvement of the AGCM in this regard is required.

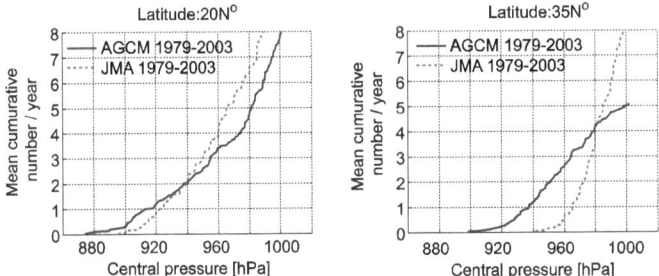

Figure 7.12: Cumulative annual average numbers of typhoons in the JMA Best track data in the period of 1979-2003 and the AGCM simulation for the current climate as a function of the central pressure at latitudes 20N ° (left) and 35N ° (right).

It should also be emphasized here that the geographical differences in the change rate of the risk over Japan may be due to patterns in the AGCM simulation results that are not physically and/or statistically significant; further investigation is suggested to verify this observation.

Modeling of wind vulnerability

The damage ratio model developed and employed in the present study relies on a narrow range of damage observation temporally (only on one typhoon event) and spatially (only on Kyushu, part of Japan). Furthermore, it only describes the relation between wind speed and the ratio of damage or probability of damage occurrence; indifferent to the degree of damage or amount of monetary losses. Here, it is suggested that research effort is directed to develop a vulnerability model based on an engineering approach; i.e. modeling of the vulnerability accommodating physical processes leading to relevant damages and their consequential losses. In the case of the assessment for Japan, an emphasis should be given on the vulnerability modeling for non-structural components of buildings such as cladding and glazing, since these are considered to be the major part of the damages and losses due to strong wind in Japan.

7.2.5. Surface roughness and development of the society

Finally, but not least, the development of the human society over time should have significant impact on the change of the typhoon wind risk. Firstly, it should affect the future development of the climate change through the general consumption behavior of the society; this concerns the scenarios in the IPCC report. Secondly, the development of the human society may change the pattern of the land use and building density, which can result in different surface roughness. In addition to these, for a comprehensive assessment of the impact, other changes of the societal structure such as population and geographical distribution thereof must be taken into account.

7.2.6. Acknowledgements

I also would like to thank Professor Takashi Maruyama for the cooperation in this study on climate change. He provided the output of the climate model and a vulnerability model, which were used as a basis for this study. This work was conducted under the framework of the "Projection of the change in future weather extremes using super-high-resolution atmospheric models" supported by the KAKUSHIN Program of the Ministry of Education, Culture, Sports, Science, and Technology (MEXT). The calculations of the numerical simulations of the mesoscale meteorological model JMA-NHM were performed on the Earth Simulator. The author thanks to Mr. Murakami for providing the datasets of the typhoon events extracted from the numerical simulations.

8. Application: Risk assessment of a approaching typhoon and real-time decision making

This application shows how the Bayesian framework for probabilistic modelling of typhoon risks can be used for the risk assessment and for real time decision making in the case of an approaching typhoon by conditioning the typhoon model with new available information.

If a typhoon has just emerged and is approaching the Japanese Islands, the typhoon model described in Chapter 2 can be used to simulate possible developments of this new typhoon by using the available data of the approaching typhoon as initial conditions for the typhoon model. To perform a predictive risk analysis for a approaching typhoon the described approach is implemented in the software tool *TRAST* (described in Section 6.4). Section 8.1 describes the functions of *TRAST* which can be used to estimate the portfolio risk for a approaching typhoon.

Another application concerns real-time decision making e.g. in the context of evacuation of people and shut-down of operation of engineered facilities in the face of an approaching typhoon, see (Nishijima et al., 2009). A framework for real-time decision analysis and a example is provided in Sections 8.2 to 8.5.

For this, the feature that the entire life of a typhoon can be modelled is useful for simulating possible tracks and changes of the intensity of the approaching typhoon. Furthermore, the consideration of the SST and the seasonal differences of probabilistic characteristics of typhoon events enables utilizing additional information such as the current SST around the location of the typhoon and current season; consequently, the uncertainties associated with the transition of the typhoon can be reduced and decisions may be made more precisely.

8.1. *TRAST*: Risk assessment of a approaching typhoon

The software tool *TRAST* (described in Section 6.4) gives the option to perform a risk analysis for the case that a new typhoon is occurred. There are two different cases if the complete typhoon track is available or if the typhoon is just approaching the Japanese Islands and only a partial typhoon track is available. The two cases are described in the following sections.

8.1.1. Scenario-based simulation

For the case that a risk analysis should be performed for a typhoon event which just has passed the Japanese Islands the software tool *TRAST* provides a option to make a "scenario-based simulation". This option provides the possibility to import the typhoon track data of the new typhoon. Then the typhoon track is used in combination with the wind field model (Section 2.9), the surface friction model (Section 2.10) and the vulnerability model (Section 6.1) to perform a risk analysis using this new typhoon track data.

8.1.2. Conditional simulation

For the case that a typhoon has just occurred and is approaching the Japanese Islands, the software tool *TRAST* provides the option of making a "conditional simulation". The partial track data of the approaching

8. Application: Risk assessment of a approaching typhoon and real-time decision making

typhoon can be imported in the interface of *TRAST* shown in Figure 8.1. The partial typhoon track data is used to set the initial condition of the typhoon model (Chapter 2) which is used to simulate a selected number of possible complete typhoons. These typhoons are used in combination with the vulnerability model (Section 6.1) to perform the risk analysis.

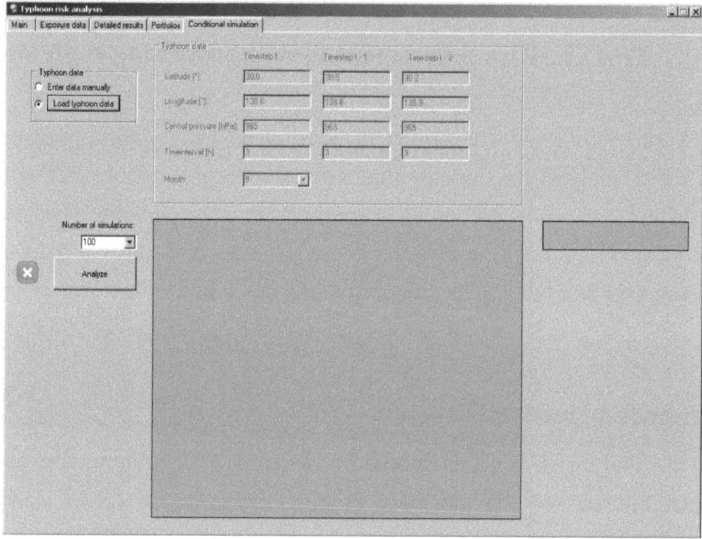

Figure 8.1: Interface for the option "conditional simulations".

8. Application: Risk assessment of a approaching typhoon and real-time decision making

The summary of the analysis results includes the following statistics (see Figure 8.2):

- Probability density function of the insured loss (①)
- Expected value of the insured losses (②)

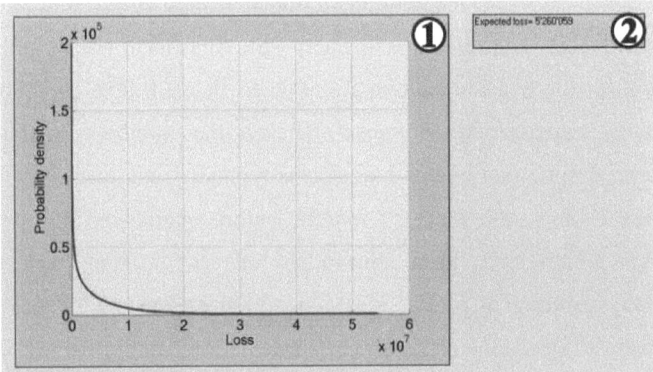

Figure 8.2: Summary of the analysis results for the case of the analysis with the option "conditional simulation".

8.2. Real time decision making

Building structures, infrastructure systems and industrial facilities (hereafter jointly referred to as engineered facilities) are often built and operated on locations where natural hazards may take place; implying significant risks. As part of the overall strategy of risk management for such engineered facilities the optional decision to shut down operation and to evacuate people and assets in the face of an emerging hazard event plays an important role. Important examples where such strategies are

8. Application: Risk assessment of a approaching typhoon and real-time decision making

presently utilized include refineries and fixed offshore platforms subject to tropical storms, storm surges and tsunamis, but also urban habitats and public infrastructure subject to events such as storms, floods, landslides, avalanches and volcanic eruptions.

The decision to evacuate a highly populated area or to shut down an industrial facility is in itself associated with significant risks and usually with high costs; evacuation operations even in well developed societies have been experienced to cause considerable numbers of fatalities. Moreover, despite that events like storms typically only last a few days it may take even in the order of months before engineered facilities like refineries which have been shut down can be brought back to full operation, resulting in production interruption and monetary losses as a consequence. Decisions on emergency evacuation, shut-down and general loss reduction activities in the face of emerging hazards are thus critical and it is of high importance from the perspective of both societal and private organizations to ensure that they are conforming to the preferences of stakeholders in regard to safeguarding life, environment and other assets. The relatively few but highly important decisions which are to be taken by just a few persons within a small time frame subject to the uncertain and incomplete information prevailing such situations must thus be well prepared.

The main challenge is to establish the theoretical and methodical basis for supporting such decisions and to derive criteria for commencement of actions in consistency with the best available knowledge on the hazards, potential consequences, the efficiency of available relevant options for risk reduction and the preferences of the stakeholders.

8.3. Characterization of the problem

The decision situations outlined above share important characteristics: (a) the hazard processes evolve relatively slow (e.g. storms and floods) and allow for reactive decision making; (b) various indicators can be observed prior to the impact of the hazard which contain information in regard to its severity (e.g. landslides, avalanches and volcanic eruptions); (c) decision makers have options for reducing risks which may be commenced at any time supported by the observed indicators. Here the typical problem arises that waiting will imply more information but might also reduce available time for evacuation and other loss reduction activities; (d) the decision making is subject to uncertainties, a part of which might be reduced at a cost; (e) on top of all, the decisions must be made fast, in near-real-time.

The present section addresses the decision problem outlined in the foregoing first in general terms. A general formulation for the optimization of decisions concerning shut-down and evacuation is presented based on the Bayesian pre-posterior decision analysis, assuming that probabilistic models are available for the assessment of the temporal and spatial evolution of emerging natural hazard events. Therein, the requirements to the probabilistic models are identified. Then, an example is considered; a decision maker, who is responsible for the operation of an offshore platform, has to decide if the operation is continued or shut-down in the face of an evolving typhoon event. The procedure for solving the decision problem using the framework is accounted for especially focusing on computational feasibility. In doing so, general technical issues involved in the decision problem formulated in the present section are identified, which are addressed as future tasks.

8.Application: Risk assessment of a approaching typhoon and real-time decision making

8.4. Decision framework

The typical time frame for decision making considered in this section range from hours to days. Within this time frame decision makers may be able to obtain information at time intervals ranging from minutes to hours. For instance, for decision making in the face of an evolving typhoon the time frame is typically several days up to a few weeks. The time interval for acquiring information on the state of a typhoon is often in the range of one to six hours as typically being the standard of meteorological agencies.

In order to facilitate decision making in situations with such relatively small time frames, it is important that relevant information available in near-real-time can be directly implemented into probabilistic models for the updating of the variables influencing underlying decision problems. Furthermore, it is necessary to establish the algorithms for identifying the optimal decision rules which apply in dependency of any potentially available information. For these purposes, a decision framework is formulated employing conditional probability representations and the sequential decision theory (De Groot, 1970), which is considered as a variant of the pre-posterior analysis (Raiffa and Schlaifer, 1961).

8.4.1. Conditional probability representations

Denote by X a random variable of significance for consequences in a decision problem, also referred to as the hazard index such as e.g. the maximum wind speed during a storm event, or the time remaining until the wind speed exceeds a critical threshold. Let $\mathbf{Y}=(Y_1,Y_2,...,Y_n)$ be a set of

8.Application: Risk assessment of a approaching typhoon and real-time decision making

random variables that characterize the phenomenon of interest and are required to calculate the probabilistic characteristics of X. Note that the subscript represents discrete times. However, the discrete times do not necessarily correspond to physical times; they may also represent the number of successive measurements undertaken for collecting information. n is the number of the times when the decisions are made, and a terminal decision (the definition is provided later) must be made before or at the nth time. Denote by $E = (E_1, E_2, ..., E_n)$ a set of random variables that represents information available at respective times, which may be utilized to reduce the uncertainty associated with the components of $\mathbf{Y}$ and in turn the hazard index X.

A simplistic but typical relationship between the variables is shown in Figure 8.3. In the figure, each node represents a variable, and each directed arrow represents the probabilistic characteristics between the variables connected by the arrow. For instance, the arrow directed from the node Y_1 to the node E_1 represents that the random variable E_1 is characterized by the conditional probability $P[E_1 | Y_1]$. When more than two arrows are directed to a node, it signifies that the random variable represented by the node is characterized by the conditional probability on the variables represented by the nodes from which the arrows are directed.

When all the conditional probabilities and the (unconditional) probabilities for the nodes to which no arrow is directed are given, conditional probabilities of any variables in the graph can be calculated. Especially, it is important in the decision framework considered here that the conditional probability of X given information $E_1, E_2, ..., E_i$ ($i = 1, 2, ..., n$) and the conditional probabilities of E_{i+1} given $E_1, E_2, ..., E_i$ ($i = 1, 2, ..., n-1$) can be easily calculated.

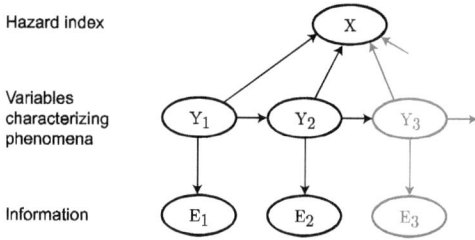

Figure 8.3: Probabilistic model representation.

Such probabilistic models and the algorithms for calculating the conditional probabilities are assumed to be available for the decision problems considered. In fact, Bayesian Probabilistic Networks are suitable for such conditional probability representation; generic algorithms are available for calculating conditional probabilities, see e.g. (Jensen and Nielsen, 2007) . However, other conditional probability representation such as regression models may be suitable as are employed in Section 8.5.

8.4.2. Decision optimization

One of the characteristics of the decision problem considered in the present chapter is that the decision maker has an option to postpone the decision. The decision maker can use the time gained by postponing the decision for reducing the uncertainties concerning the decision problem. There two possible approaches to reduce the uncertainties are possible.

One approach is associated with the reduction of the aleatory uncertainty involved in the decision problem; that is, as a function of time the temporal and spatial frame of the phenomenon underlying the decision problem change in such a way that the uncertainty of the variable of direct interest for the decision making (e.g. the hazard index X in Figure 8.3) is reduced, see (Faber, 2005). Thus, by postponing the decision the probability that the decision maker makes a suboptimal decision can be reduced. Note, however, that the probability is also increased that loss reduction activities are undertaken too late (if they are necessary). The other approach to reduce the uncertainty is by means of reducing the epistemic uncertainty associated with the underlying phenomenon. This is possible by collecting more information through e.g. monitoring and in-situ surveys. In this way, too, the probability that the decision maker makes a suboptimal decision can be reduced. However, such measures for collecting more information are worth undertaking only if the costs for collecting more information are smaller than the expected value of the additional costs arising from the suboptimal decision, which otherwise could be avoided. Thus, it is essential in near-real-time decision making to identify the extent to which the decision maker shall reduce the uncertainty by investing the time (i.e. postponing the decision) and costs for collecting more information before making the decision.

For convenience, the following terminology is introduced: a "terminal" decision refers to the decision which must eventually be made in a given decision problem, and decision making terminates when a terminal decision is made; a decision to "postpone" refers to the decision that a terminal decision is not made and, meanwhile, the decision maker collects more information. The decision maker can choose to "postpone" as many times as she/he needs.

In the following sections, the decision framework is presented. Two decision situations are differentiated; one-time decision making and sequential decision making. The former forms a building block for the latter, which is the main focus of the present chapter. Note that whereas in the formulation of the decision problem in the following the random variables are assumed to be continuous, a similar discussion holds for the case of discrete variables and the mixture of both.

8.4.3. One-time decision making

In one-time decision making, a decision maker needs to make a terminal decision out of a set of possible decision alternatives $A = \{a_1, a_2, ..., a_m\}$ without postponing the decision. The ingredients necessary for identifying an optimal decision are the conditional probability $f(x|e)$ of the hazard index X given the information e, and the utility function $U(x,a)$ as a function of the outcome x of X and $a \in A$. The optimal decision a^* is determined as:

$$a^* = \arg\max_i E[U(X, a_i) | e] \tag{7.1}$$

namely, the optimal decision a^* is the decision that maximizes the expected utility given the information e. This is often called as posterior decision analysis.

8.4.4. Sequential decision making

In sequential decision making, another decision alternative a_0, i.e. postponing the terminal decision, is introduced which facilitates for collecting and assessing more information before the terminal decision is made. The additional ingredients necessary for identifying the optimal decision are the conditional probability $f(e_{t+1}|e_1,...,e_t)$ of e_{t+1} given the information $e_1,...,e_t$, and the conditional expected utility $E[U(X,a)|e_1,...,e_t]$ given information $e_1,...,e_t$ for $a=a_1,a_2,...,a_m$. The subscript t is introduced to distinguish the information at different times (see Figure 8.3). Employing the concept of sequential decision theory, the optimal decision $a^*(e_1,...,e_t)$ given information $e_1,...,e_t$ at time t is obtained as the one that maximizes the expected utility $E[U(X,a)|e_1,...,e_t]$:

$$E[U(X,a^*(e_1,...,e_t))|e_1,...,e_t] = \max_{i=0,1,...,m} E[U(X,a_i)|e_1,...,e_t] \text{ for } t=1,2,..,n-1 \quad (7.2)$$

$$E[U(X,a^*(e_1,...,e_t))|e_1,...,e_n] = \max_{i=1,...,m} E[U(X,a_i)|e_1,...,e_n] \text{ for } t=n \quad (7.3)$$

where

$$\begin{aligned} E[U(X,a_0)|e_1,...,e_t] &= \int E[U(X,a^*(e_1,...,e_{t+1}))|e_1,...,e_{t+1}] \\ &\cdot f(e_{t+1}|e_1,...,e_t)de_{t+1} \text{ for } t=1,2,...,n-1 \end{aligned} \quad (7.4)$$

Equation (7.4) reflects that the expected utility at time t given that the decision is postponed is equal to the expected value of the maximized expected utility at the next time; the optimal decision at time $t+1$ must be known prior to identifying the optimal decision at time t. Thus, solving the equation system defined by Equations (7.2)-(7.4) requires backward calculation in general. A simplistic numerical example for this and also another way for solving the equation system are provided in (Nishijima et al., 2008a). A practical way for solving the equation system will be explained along with the example in the next section.

8.5. Example

8.5.1. Problem setting

A decision maker is requested to make the decision on whether or not the operation of an offshore platform should be shut down in the emergence of a typhoon event. The possible decision alternatives are the terminal decisions of shut-down (a_1), no shut-down (a_2) and postponing the terminal decision (a_0). When the decision maker chooses a_0, she/he can obtain further information on the typhoon such as the position, central pressure, translation speed and angle of the typhoon. The information is assumed to be provided by a meteorological agency every six hour, which incurs no cost. It is assumed that the completion for shutting down the operation takes twelve hours after making the terminal decision a_1.

In what follows, the assumptions made and models employed are explained.

8.5.2. Typhoon model

For modelling the wind speed loaded on the platform due to the typhoon, the typhoon model is employed which is developed by the author, see (Graf et al., 2009) and (Nishijima and Faber, 2007) for the overview and relevant literature survey. The typhoon model is composed of five components; occurrence model, transition model, wind field model, surface friction model and vulnerability model. For the example considered here, only the transition model, wind field model and surface friction model are of relevance. For simplicity, the filling model (a part of the transition model) that describes the change of the central pressure of typhoons after making landfall is not considered in the example. The models are described in Chapter 2. Thus, using the models describe, the wind speeds at the platform can be calculated given the transition states of the typhoon.

8.5.3. Postulated consequence model

The platform is assumed to be damaged only if the wind speed u exceeds the threshold u_c ($=38[m/s]$) while the platform is in operation (the expected damage cost, $C_D = 10$); the platform is assumed not to be damaged if the wind speed does not exceeds the threshold, or if the platform is successfully shut-down (i.e. not in operation when the wind speed exceeds the threshold). However, in the latter case the cost for production interruption must be considered (the expected production interruption cost, $C_{PI} = 1$). Three cases are possible for which the expected damage costs C_D are incurred; the first is the case where the decision a_2 is made and the wind speed exceeds the threshold, the second is the case where the decision a_1 is made but the wind speed exceeds the threshold before the shut-down is completed and the last is the case where the decision a_0 is made and the wind speed exceeds the threshold before the next time when the decision is made. No consequence occurs if and only if the decision a_2 is made and the wind speed does not exceed the threshold. The expected costs C_{PI} for production interruption is incurred if and only if the decision a_1 is made and the shut-down is completed before the wind speed exceeds the threshold (if it does), or the wind speed does not exceed the threshold.

Table 8-1: Conditions and associated losses postulated in the consequence model.

Facility	Wind speed	
	$u \geq u_c = 38[m/s]$	$u < u_c = 38[m/s]$
In operation	$C_D = 10$	0
Not in operation	$C_{PI} = 1$	$C_{PI} = 1$

8.5.4. Other conditions

For illustration purposes, it is assumed that the initial transition states of the typhoon are known and the radius of maximum wind speed R_M is known and is constant during the typhoon event. Table 8-2 summarizes the assumed initial conditions. The location of the platform is also shown.

8. Application: Risk assessment of a approaching typhoon and real-time decision making

Table 8-2: Assumed initial conditions.

Central pressures at $t=-1,0$ and 1	930, 930 and 930 [hPa]
Translation speeds at $t=1$	20[km/h]
Translation angles at $t=0$ and 1	0[°] and 0[°] (Northwards)
Position at $t=1$	(129°E, 28°N)
Sea surface temperature at the location of the typhoon at $t=1$	27.9 [°C]
Radius of maximum wind speed, R_M	100 [km]
Location of the platform	(130°E, 31°N)

8.5.5. Algorithm

In this decision problem, it is assumed that the decision is terminated within 36 hours. The time frame is discretized by the time interval of six hours; i.e., the time frame is discretized into six time steps where the information becomes available and the decisions are made. This assumption seems reasonable since the typhoon is very likely to pass though the area relevant for the facility at the 6th time step, see Figure 8.4. The figure shows two possible transition (indicated by dashed lines with circles) of the typhoon in order to facilitate an understanding of the decision situation considered in the example. For each discretized time, the space of the possible states of the typhoon is discretized into $3^3=27$ states (3 states for the central pressure, the translation speed and the angle respectively) in an adaptive manner; that is, the space of the states at the different times is discretized in such a way that the probability of all states is equal to 1/27; since the conditional probability distributions of the states at the times differ depending on the previous states the discretized states at

8.Application: Risk assessment of a approaching typhoon and real-time decision making

each time differ in the different branches in the event/decision tree show in Figure 8.5.

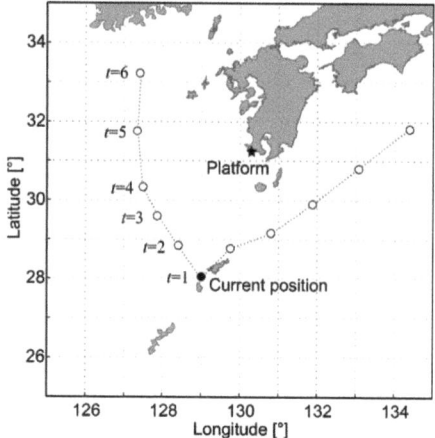

Figure 8.4: Illustration of the transition of the typhoon and the location of the platform.

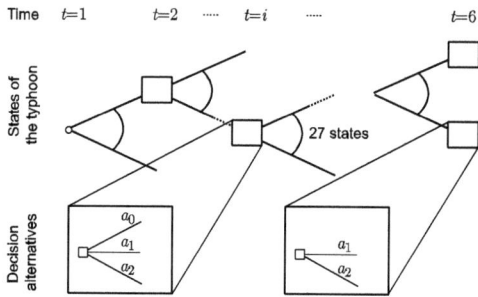

Figure 8.5: Event/decision tree of the decision problem considered in the example.

The information E_t introduced in the formulation of the decision framework corresponds to the (discretized) states of the typhoon in this

8.Application: Risk assessment of a approaching typhoon and real-time decision making

example. Note that the capital symbol E_t represents that E_t is a random variable and the symbol e_t (used in the following) represents the realization of the random variable E_t. The set of the information $(e_1,...,e_6)$ can be used to identify a specific branch in the event/decision tree, and the information $(e_1,...,e_t)$ up to the time t can be used to identify a set of branches which share the same branch up to the time t. Note in the event/decision tree all the branches are assumed to have the same length (i.e. n time steps); however, the branches that have more than two terminal decisions are valid only up to the time step when the first terminal decision is made. The remaining parts of the branches are referenced only to calculate the probabilities that the wind speed exceeds the threshold in the remaining time. Furthermore, the branches in which the wind speed exceeds the threshold are valid only up to the time when the wind speed exceeds the threshold, and thus the remaining part of the braches should be regarded as dummies. In solving the decision problem, these parts of the branches are never used.

The procedure for identifying the optimal decisions for the given information at each time step is shown in the following. Consider the branches indentified by the set of the information $(e_1,...,e_5)$, i.e. sharing the same branch up to the 5th time step and having different branches between the 5th and 6th time steps. Check if the parts of the branches between the 5th and 6th time steps are valid (whether or not the wind speed exceeds the threshold and the terminal decision is made at earlier time steps). For this it is necessary to calculate the wind speeds at the time steps $t = 1, 2,..., 5$

8. Application: Risk assessment of a approaching typhoon and real-time decision making

for this branch, which are denoted by $u_1(e_1)$, $u_2(e_1,e_2)$, $u_3(e_1,e_2,e_3)$, $u_4(e_1,...,e_4)$ and $u_5(e_1,...,e_5)$ respectively. Also, calculate the wind speeds $u_6(e_1,...,e_6)$ at the 6th time step for all possible e_6, and the probability that the wind speed exceeds the threshold at the 6th time step, which is denoted by $p_6(e_1,...,e_5)$ — these are used later. Note that the wind speed associated with each time step is the maximum of the wind speeds calculated by 10-minute time interval between the subsequent time steps (for instance $u_2(e_1)$ is the maximum wind speed between the first and the second time steps); thus, the situations where the maximum wind speed is realized between the time steps are considered.

If the parts of the branches are valid, the optimal decisions at the 6th time step for the given set of the information $(e_1,...,e_5)$ and all possible information e_6 are obtained in accordance with Equation (7.3). Here, the utility is understood as the negative of the cost, and instead of the maximization of the expected utility the minimization of the expected value of the costs is considered. Thus, the optimal decision $a^*(e_1,...,e_5)$ at the 5th time step for this branch and the minimized expected cost $E[U(X,a^*(e_1,...,e_5)) | e_1,...,e_5]$ is obtained in accordance with Equation (7.4). Note that in the case when the wind speed exceeds the threshold between the 4th and 5th time step, the decision alternatives at the 5th time step are indifferent, and all resulting in the damage of the facility, thus the expected costs are C_D.

8. Application: Risk assessment of a approaching typhoon and real-time decision making

Repeat this procedure for different information e_5, but for the same set of the information $(e_1,...,e_4)$. In doing so, the wind speeds $u_1(e_1)$, $u_2(e_1,e_2)$, $u_3(e_1,e_2,e_3)$ and $u_4(e_1,...,e_4)$ can be reused, but the wind speed $u_5(e_1,...,e_5)$ must be recalculated. Also, calculate the probabilities that the wind speed exceeds the threshold at the 5th and 6th time steps, which are denoted by $p_5(e_1,...,e_4)$ and $p_6(e_1,...,e_4)$ respectively, whereby the probability $p_6(e_1,...,e_5)$ for different e_5 can be used to calculate these probabilities. Once the probabilities $p_5(e_1,...,e_4)$ and $p_6(e_1,...,e_4)$ are calculated, the probabilities $p_6(e_1,...,e_5)$ are not necessary for further calculation. Then, it is possible to identify the optimal decision $a^*(e_1,...,e_4)$ at the 4th time step for this branch, and the minimized expected cost $E[U(X,a^*(e_1,...,e_4)) | e_1,...,e_4]$; the expected cost for the decision a_0 is calculated using the minimized expected costs $E[U(X,a^*(e_1,...,e_5)) | e_1,...,e_5]$ and the conditional probabilities $f(e_5 | e_4)$ for all possible information e_5; the expected costs for the decisions a_1 and a_2 are directly calculated using the probabilities $p_5(e_1,...,e_4)$ and $p_6(e_1,...,e_4)$. Once the minimized expected cost $E[U(X,a^*(e_1,...,e_4)) | e_1,...,e_4]$ at the 4th time step is obtained, the minimized

8.Application: Risk assessment of a approaching typhoon and real-time decision making

expected costs $E[U(X,a^*(e_1,...,e_5)) | e_1,...,e_5]$ at the 5th time step are not necessary for further calculation.

Repeat this procedure for different information e_4 but for the same information (e_1,e_2,e_3). Then, the optimal decision and the minimized expected cost at the 3rd time step are obtained. By repeating this procedure, the optimal decision $a^*(e_1)$ is finally obtained.

8.5.6. Results

Following the algorithm described above the optimal decision at the initial time is identified. The identified optimal decision $a^*(e_1)$ is to "postpone" (a_0), whereby the calculated expected costs for the decision alternatives a_0, a_1 and a_2 are 1.61, 1.68 and 1.63 respectively.

Assume that the decision maker chooses a_0 at the initial time and then the typhoon moves to the location (129°E, 28.5°N) with the central pressure of 925 [hPa] and the sea surface temperature of 28.0 [°C] in the next six hours. The optimal decision at this time can be identified by applying the algorithm once again but including the new information. Note that in the application of the algorithm at this time the initial time is reset to the current time (i.e. six hours later than the initial time set in the previous optimization) and the optimization is performed. The intermediate calculation results in the previous optimization cannot be reused because these results are not stored as explained in the previous section. Even though the results are stored these may not be useful directly useful since it is likely that the new information corresponds to none of the 27 discretized states. The optimal decision thus identified is to "shut-down" (a_1), whereby the calculated expected costs for the decision alternatives a_0, a_1 and a_2 are 2.77, 1.88 and 3.02 respectively. In this way, with the decision framework presented in this chapter it is possible to identify the optimal decisions as a function of the information that becomes available in near-real-time.

8.5.7. Discussions

Whereas the decision framework presented in this chapter is general and in principle can be applied to a broader range of practical decision problems characterized as described in Section 8.3, there is a need to consider computational capability in practice; for instance, in solving the decision problem considered in the example, the space of the possible states at each time step is discretized into 3 for each variable. Although this is a coarse discretization, a finer discretization would prohibitively increase the calculation time required for solving the problem.

The calculation time necessary for identifying the optimal decisions is approximately proportional to m^n, where m is the number of possible states at each time step and n is the number of discrete times at which decisions can be made. A practical solution for reducing the calculation time is to reduce the number of the times. However, this can easily lead to suboptimal solution especially for the type of the decision problems considered in the present chapter, i.e. decision problems where "postpone" is a decision alternative. This is because the decision could be forced to

8.Application: Risk assessment of a approaching typhoon and real-time decision making

terminate at the earlier time step where the optimal decision might be to "postpone" if the number of considered times is smaller.

The following approaches may be useful for the reduction of the calculation time, whereby avoiding the problem mentioned above. One approach is to start the decision analysis at the time when the terminal decisions are likely to be optimal decisions. That is, the times when to "postpone" is dominantly the optimal decision are eliminated from the time frame of the decision analysis. Another approach is to split the time frame of the decision analysis into different lengths. In this way, the duration of the decision time frame can be sufficiently long, but the number of the time steps can be reduced. However, in each respective approach the choices of the appropriate time frame and time intervals would require trial-and-errors. This issue is addressed as a future task.

In the example the intermediate calculation results in the decision optimization for the given initial condition are not stored and the calculation for the decision optimization is performed when the new information becomes available. However, it is possible to store these intermediate results and to use them for identifying the optimal decisions at the subsequent times by interpolating the calculated expected costs for different decision alternatives and different potentially available information, even though the information factually available does not correspond to any discretized states. Furthermore, in principle it is possible to calculate the expected costs for all possible decision alternatives and all relevant conditions prior to the emergence of natural hazards, to store these and to use for identifying the optimal decision when

8. Application: Risk assessment of a approaching typhoon and real-time decision making

the hazard events occurs; however, the amount of the data which have to be stored may be large.

The calculation times needed to identify the optimal decisions in the example are up to several hours with a standard PC depending on the initial conditions. Thus, if the calculations for the decision optimizations are decided to make in near-real-time, these calculations must be made with several PCs in parallel assuming several initial conditions that are likely to occur at the next time when the decision is made.

9. Conclusions and outlook

9.1. Summary

In this thesis a Bayesian framework for the probabilistic modelling of typhoon risks is proposed. This includes: (1) A probabilistic typhoon model which satisfies the requirements of the proposed framework. (2) The means to consider the uncertainties in the modelling of the typhoons and in the typhoon risk analysis. (3) The means to update and condition the typhoon model by new available data and information.

The developed Bayesian framework can incorporate all relevant information available. Using this information enables a more precise risk assessment in two different ways. (1) For a risk assessment of an approaching typhoon event actual information can be used to condition the model. This reduces the uncertainties for a specific risk analysis. (2) By using the information to update the parameters in the model, the overall modelling uncertainties can be reduced over time.

The Bayesian framework does not only include the methodology to establish a typhoon model but also the mechanisms which enable the decision makers to take into account the available information during the process of decision making and thus facilitate the updating of the parameters in the model over time. Three main features are included in the framework: (1) estimation of annual average loss/probable maximum loss; (2) estimation of loss of any given portfolio when a typhoon event has initiated and is approaching the considered region; (3) updating of the

9. Conclusions and outlook

models with all the data available after one or more typhoon events have occurred.

The proposed Bayesian framework is applied to the region of Japan and a typhoon risk model is established with the focus on the following features which facilitates to incorporate the available information: (1) typhoon events are modelled for the entire life of typhoons, i.e. from occurrence to dissipation; (2) the effects of sea surface temperature (hereafter, SST) on the evolution of typhoon events are accounted for; (3) seasonal differences of the probabilistic characteristics of the transition of typhoons are accounted for. The hazard model is composed of sub-models, describing all phases of the typhoon hazard process starting with the occurrence of typhoons over the spatial and temporal development of typhoons including landfall and possible filling and ending with the probabilistic characterization of extreme wind speeds at any location in Japan.

The application of the Bayesian framework for the probabilistic modelling of typhoon risk is illustrated in three examples: (1) Combining the hazard model with a vulnerability model (which represents the probability distribution of the loss of individual exposures as a function of the wind speed) enables the risk assessment of insurance portfolios. The use of the framework for the estimation of insurance portfolio risk analysis shows how the uncertainties are considered for the risk estimation. (2) Considering the entire life of typhoons and the effects of the SST on the evolution of the typhoons enables the investigation of the effects of global warming on the probabilistic characteristics of strong wind speed induced by typhoons. The assessment of the effect of global warming shows how the framework can be updated with new available data. (3) The third

9. Conclusions and outlook

application concerns real-time decision making e.g. in the context of evacuation of people and shut-down of operation of engineered facilities in the face of an approaching typhoon. For this, the feature that the entire life of a typhoon can be modelled is useful for simulating possible tracks and changes in the intensity of the approaching typhoon. Furthermore, the consideration of the SST and the seasonal differences of probabilistic characteristics of typhoon events enables utilizing additional information such as the current SST around the location of the typhoon and current season; consequently, the uncertainties associated with the transition of the typhoon can be reduced and decisions may be made more precisely. The application for real time decision making in the case of a approaching typhoon shows how the framework can be conditioned by new available information.

9.2. Conclusions

This thesis proposes a Bayesian framework for the probabilistic modelling of typhoon risks. The presented Bayesian framework is a first step towards a full probabilistic treatment of typhoon risk analysis. The Bayesian framework includes: (1) A probabilistic typhoon model which satisfies the requirements for the proposed framework. The probabilistic typhoon model is successfully developed and the verification shows a good agreement with the historical data. (2) The consideration of the uncertainties involved in the modelling of the typhoons and in the portfolio risk analysis and the statistical uncertainties due to parameter estimation. A framework to include the uncertainties due to the selection of the models and the assumptions made in the typhoon model, is presented. (3) The means to update and condition the Typhoon model with new available data and information.

The application of the Bayesian framework for the probabilistic modelling of typhoon risk is illustrated in three examples: (1) The use of the framework for the estimation of insurance portfolio risk analysis shows how the uncertainties are considered in the estimation of risks. (2) The assessment of the effects of global warming shows how the framework can be updated with new available data. (3) The application for real time decision making in the case of an approaching typhoon demonstrates how the framework can be conditioned by new available information.

In the sections below, the conclusions for the individual parts of the thesis are discussed.

9.2.1. Typhoon model

The proposed Bayesian framework for probabilistic modelling of typhoon risks includes a developed typhoon model for the region of the north west pacific. The approach employed for the development of the model, which takes basis in the state-of-the-art research works, is introduced. For the verification and validation of the employed approach and the developed model, the performance of the developed typhoon model is assessed in several ways by comparing the simulation results obtained using the developed model with historical observations. Finally, the advantages of the main features of the developed model in practical applications are emphasized and several applications of the developed mode are introduced.

For the probabilistic typhoon model the following components are established:

- A new occurrence model is developed, which represents the occurrence of a typhoon as a function of the location, the season (month) and the SST.
- For the transition model the proposed approach from (Vickery et al., 2000) is followed, but the model is adapted for the region of the North West Pacific and modified so that the development of a typhoon is also a function of the season (month) and the SST.
- The wind field model is established as proposed by (Georgiou et al., 1983).
- The surface friction model is developed as proposed by (Meng et al., 1997). A new scheme for the estimation of the roughness length is developed which combines the two approaches described in the state of the art.
- A model to estimate the portfolio losses is developed containing a new vulnerability model which considers the epistemic uncertainties.

A hazard event set builder software tool is developed. This event set builder can be used to automatically create a stochastic event set using the described typhoon model.

9.2.2. Treatment of epistemic uncertainties in the typhoon model

The proposed Bayesian framework shows how the uncertainties in the modelling of typhoon risks are treated. The epistemic uncertainties in the developed typhoon model due to the modelling of the phenomena are quantified for each sub model. The occurrence model is based on a

9. Conclusions and outlook

Bayesian network, which provides the empirical probability distribution of the frequency and the location of the occurs of typhoons. The transition model is based on a regression model and considers the modelling uncertainties in the error term of the regression. It would be possible to consider the uncertainties of the parameters of the regression model itself by assuming random variables for each parameter, but this has to be investigated further and would increase the computation time of the typhoon simulation since much more random variables has to be simulated. The wind field model and the surface friction model are deterministic models and the uncertainties associated with the deterministic wind field model and the surface friction model are considered indirectly in the vulnerability model. Since the simulation of the wind field is the most computational intense part in the typhoon simulation, it was decided to use deterministic models, but it would be possible to replace the wind field model and the surface friction model with a probabilistic model by e.g. introducing one or more random variables into the model.

The vulnerability model is established using the historical damage data provided by Aon Benfield Japan in combination with the corresponding wind speed, which in turn is estimated using historical typhoon tracks applying the wind field model and the surface friction model. The probability distribution of the loss estimation contains the uncertainties of the vulnerability model itself as also the uncertainties of the wind field model and the surface friction model.

For considering the epistemic uncertainties due to the model selection and the assumptions in the typhoon model, a framework is presented. This

9. Conclusions and outlook

framework shows how the uncertainties can be quantified and integrated in the risk assessment and the decision analysis processes.

This framework addresses three challenging issues in tropical cyclone risk management from the perspective of the treatment of uncertainty. These are:

- Separation of aleatory and epistemic uncertainties.
- Quantification of epistemic uncertainty.
- Implementation of these uncertainties in the formal framework for risk assessment and decision analysis.

As a first step for addressing these issues, a quantification of the variability of hazard assessment results due to different alternative models is investigated.

Taking basis in the developed typhoon model, eight different alternative models are systematically developed (i.e. no subjective "tuning-up"). The variability of the 100-year return period wind speed at Tokyo is approximately 3 [m/s]. It is found that the major contribution to the variability comes from the discretization schemes and the data set utilized for modelling of the alternative models.

Whereas a number of alternative models are investigated in Chapter 4.9, more models are proposed and feasible. For instance, Hall and Jewson (2007) propose a modeling scheme different from the scheme considered here; for transition modeling, they propose to consider all tropical cyclones with different weights as a function of distance, instead of discretization. Here, it should be mentioned that the consideration of more alternative models does not necessarily increase the variability, as long as the variability is quantified in a rational manner and not naively in terms of e.g. a difference between the upper bound and lower bound. For this purpose, however, the quantification of epistemic uncertainty is required. Without going into theoretical detail, it is said that the quantification of such epistemic uncertainty can be investigated within the framework of model selection and/or model weight (see e.g. (Raftery et al., 1997; Hoeting et al., 1999; Eklund and Karlsson, 2004; Ando and Tsay, 2009; Ando, 2010; Riggelsen et al., 2011)). However, as of yet, no general framework is developed, which is a challenging task not only in tropical cyclone risk management, but also in general.

9.2.3. Updating the typhoon model

The process of managing risks due to natural hazards may be considered as a repeated sequence of actions with the purpose of optimizing measures of risk reduction, collection of information and updating of models. From this perspective any risk management tool should have the potential to incorporate available data for the purpose of updating models in an efficient and consistent way. The present thesis proposes a Bayesian probabilistic modelling approach trough which random variables represented in a probabilistic network are updated or conditioned using information achieved from observations. The proposed approach is applied to three illustrative examples showing the advantages; the approach provides a relatively easy way to include new information into a probabilistic risk assessment. Finally, the possible situations where biases may occur in the process of updating probabilistic models are pointed out and possible solutions to circumvent the biases are presented.

A main feature of the proposed Bayesian framework for probabilistic modelling of typhoon risk is the updating of the models with all the data available after one or more typhoon events have occurred. This feature facilitates to update the typhoon model after a certain period of time, for example at the end of a year, when all the information is organized as data. So, over time, the model can be seen to better represent the phenomena.

For updating the developed typhoon model a model builder software tool has been created. This model builder automatically establishes a typhoon model using as input all the available information. The model builder is also used to establish a typhoon model which is based on a) only a part of the available data for the validation of the typhoon model, b) on a different functional form or data sets to assess the difference between alternative models and c) on a data set obtained from a climate model to assess the effect of global warming.

9.2.4. Portfolio risk analysis

The example of the application of the proposed Bayesian framework for the assessment of the risk of insurance portfolios shows how the uncertainties can be considered in the risk assessment.

A portfolio loss model is developed which considers the uncertainties and the correlation of the individual building losses. The proposed model frame work is able to consistently take into account dependencies in the components of the portfolio losses and thus more realistically models the risks as compared to previous models.

A vulnerability model is established based on insurance data, which considers the model uncertainty. This vulnerability model represents the relation between the wind speed induced by a typhoon and the probability that a loss occurs and the amount of loss. For the verification of the developed vulnerability model, the performance of the developed model is assessed by comparing the simulation results obtained using the developed model with historical observations. The verification shows a good agreement for strong as well as for weak typhoons.

Due to the feature that the seasonal differences of the probabilistic characteristics of typhoon events are considered, it is possible to estimate portfolio losses in a certain period in a year. This is useful in practice when the assessments of portfolio losses are required for the remaining period of a year.

To allow a user friendly assessment of a portfolio risk, a software tool with the name *TRAST* is developed. *TRAST* provides a intuitive user interface to perform a risk analysis. It uses a database of a stochastic event set, which was created using the typhoon model in combination with the vulnerability model.

9.2.5. Global warming

The proposed Bayesian framework is used to investigate the effect of global warming on structural reliability in the context of a possible increase of tropical cyclone activity.

For this purpose two studies have been conducted. In the first study a probability-based engineering approach is adopted. The approach employs the probabilistic typhoon model that is developed during this thesis. The probabilistic model for the resistance of structures is adapted from the JCSS Probabilistic Model Code. First the consistency of the probabilistic

9. Conclusions and outlook

typhoon model is verified with the results of alternative models not relying on historical data. Thereafter the suggested model is applied for assessing the change of structural reliability considering the effect of the increased SST on tropical cyclone activity. Then it is also investigated to which extent the resistance of structures must be increased in order to maintain the present level of structural reliability. Although these investigations are made for structures in the northwest Pacific region the approach adopted in the present study can be applied to other regions provided that the relevant models and data are available.

In the second study a quantitative impact assessment of the climate change on civil infrastructure is performed, taking the typhoon wind risk on residential buildings in Japan as the example. The main findings are that in the future (2075-2099) at most locations of Japan: (1) extreme wind events (10-minutes sustained wind speed >30m/s) are more likely to occur; (2) the median of the annual maximum wind speed decreases; (3) the expected number of damaged residential buildings decreases, assuming that the profile of the building portfolio remains unchanged. Based on these findings, the assumptions and inputs in the assessment are critically reviewed, thereby, needs of further efforts toward more credible assessment of the impact are identified.

9.2.6. Risk assessment of a approaching typhoon

The application of the proposed Bayesian framework for the probabilistic modelling of typhoon risk for real-time decision making shows how the framework can be conditioned with new available information.

A framework for near-real-time decision making is presented taking basis in a conditional probability representation and the sequential decision theory. The framework considers two situations one-time decision making and sequential decision making. The framework can be facilitated in the situations where decision makers have to make decisions in a near-real-time decisions problem using information that becomes available in near-real-time. The present thesis considers a class of decision problems where decision makers must make decision in near-real-time in response to the information which becomes available also in near-real-time. A typical example of this class of decision problems is the decision of the evacuation of people and assets and the shut-down of the operation of engineered facilities in the emergence of natural hazards.

First, the class of decision problems considered is characterized. Next, the decision framework for the class of decision problems is formulated. The formulated decision framework takes basis in the sequential decision theory, which is a variant of the more general theory of the pre-posterior decision analysis.

The use and advantages of the decision framework are illustrated with an example. The example considered is the decision situation where a decision maker must decide if the operation of an offshore platform is to be continued or shut-down in an emerging typhoon event. In this example, the typhoon model developed in this thesis is employed.

Finally, the computational limitations and possible approaches to avoid the limitations are briefly discussed; (1) the time frame of the decision analysis is chosen in such a way that the situations where to "postpone" is dominantly the optimal decision are excluded and (2) the time frame of the decision analysis is discretized into different lengths so that the time period when the timing of the terminal decisions becomes more crucial is discretized more finely.

9.3. Scientific achievements

The output of the present thesis provide aside a better phenomenological insight of typhoon events in probabilistic terms, a Bayesian framework within which all the information and data can be rationally taken into account in the process of loss estimation. The proposed Bayesian framework is a first step to a full probabilistic treatment of typhoon risks and shows how the involved uncertainties can be consistently considered and provides the mean to update and condition the typhoon model with new available data and information. Considering consistently the involved uncertainties allow a more realistic assessment of the typhoon risk. Updating the model with new available data facilitates that over time, the model can better represent the phenomena. Conditioning the model with new available information allows a more precise risk analysis for give situation e.g. for a new approaching typhoon.

9. Conclusions and outlook

Compared to previous typhoon models the presented Bayesian framework for the probabilistic modelling of typhoon risks is developed with the scope of applying the model for a broader range of decision situations. Such decision situations include: real-time decision making for the evacuation of people and shut-down of engineered facilities in the face of emerging typhoon events; adaptation of building codes in regard to wind loads to the possible increase of wind hazards that might be caused by global climate change.

The usefulness of the proposed Bayesian framework is demonstrated in three examples. The application to insurance portfolio risk analysis shows how the uncertainties and the correlation between the individual losses can be considered. The second application shows how the Bayesian framework can be used to assess the effect of a climate change by conditioning and updating the model with new available information and data. The third application demonstrates how the proposed Bayesian framework can be used for real-time decision making in the case of a approaching typhoon by conditioning the models with the new available information.

The proposed framework is implemented in three software tools. A model builder software tool is developed, which enables to establish and update a typhoon model using all available information. A hazard event set creator software tool is developed, which establishes a data base of stochastic typhoon events using the typhoon model to simulate typhoon events and the software tool *TRAST* (Typhoon Risk Analysis Software Tool), which provides a user friendly graphical interface to perform a risk analysis for a insurance portfolio. The present tools have the following advanced features: The uncertainties are consistently quantified, the uncertainties can be reduced by conditioning the models by incorporating all available information, over time the phenomena is better represented by updating the models with new available data. The implemented software tools support decision makers in practical applications and enhance the decision making.

9.4. Outlook

The research presented in this thesis contains several topics, which should be further investigated and which present several opportunities for follow-up research projects. These topics are summarized in this section.

The typhoon model is developed for the region of the North West Pacific and verified for the Japanese Islands. The presented typhoon model can be easily adapted to other countries in this region. Apart from the verification of the model for the selected country, only the surface friction model has to be re-established using the land use data and the topological maps of the considered country. Having a detailed model for different countries in the region of the North West Pacific would enable to assess the correlation of the losses in different countries due to the same typhoons. The presented framework is generic in the sense that it is formulated in terms of observable indicators and can thus be easily be implemented for the characteristics of a different region e.g. the Atlantic basin.

A framework for the treatment of the epistemic uncertainties due to model selection is proposed in this thesis. An example shows how the hazard variability due to different alternative models can be assessed. A remaining task is to find an appropriate formulation to define a weight, which represents the degree of belief, for each alternative model to be able to combine the different alternative models and to integrate the uncertainties due to the model selection into the risk assessment.

9. Conclusions and outlook

The application of the Bayesian framework for risk assessment of a approaching typhoon showed that the problem has to be simplified to avoid the computational limitations. To overcome these computational limitations and to develop a framework how real time decisions based on the typhoon model can be made, Annett Anders and Kazuyoshi Nishijima are working on a research project at DTU with the title: Real Time Decision Support in the Face of Evolving Natural Hazards (Anders and Nishijima, 2011).

Other hazard sources associated with typhoons such as floods and storm surges are presently not considered and the modelling of these hazard sources is addressed as a future task. The wind speed is used as a hazard indicator in the typhoon model, however several losses cannot be explained by the wind speed alone. Several losses occur due to floods caused by typhoons. As a first step to establish a flood model, a precipitation model is currently under development. This precipitation model simulates the number and the location of rain cells and the amount of precipitation during the lifetime of a typhoon. Rocco Custer and Kazuyoshi Nishijima are currently working on a research project at DTU to develop a hazard risk model framework with application to flood risk, based on the typhoon model and this precipitation model.

10. Appendix

10.1. Appendix A - Verification of the transition model

As described in Section 3.2, Figures 10.1 to 10.3 show for the month July, August and September the simulation results of the transitions of typhoons compared with the historical observations with respect to: The frequency of the typhoons with certain intensities which cross certain latitudes (20°, 25°, 30°, 35° and 40°), translation angle and translation speed of the typhoons at the moment when the typhoons cross these latitudes between the longitudes [120°, 160°].

10. Appendix

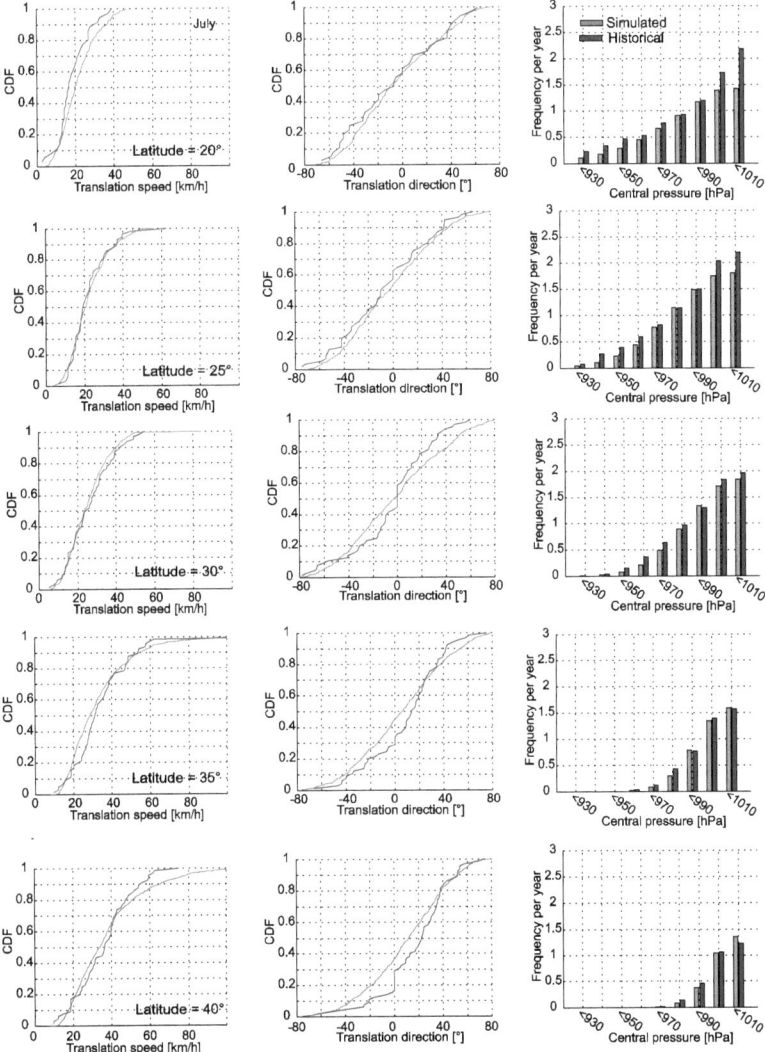

Figure 10.1: Cumulative distributions of the translation speed and direction and central pressure of typhoons in July.

10. Appendix

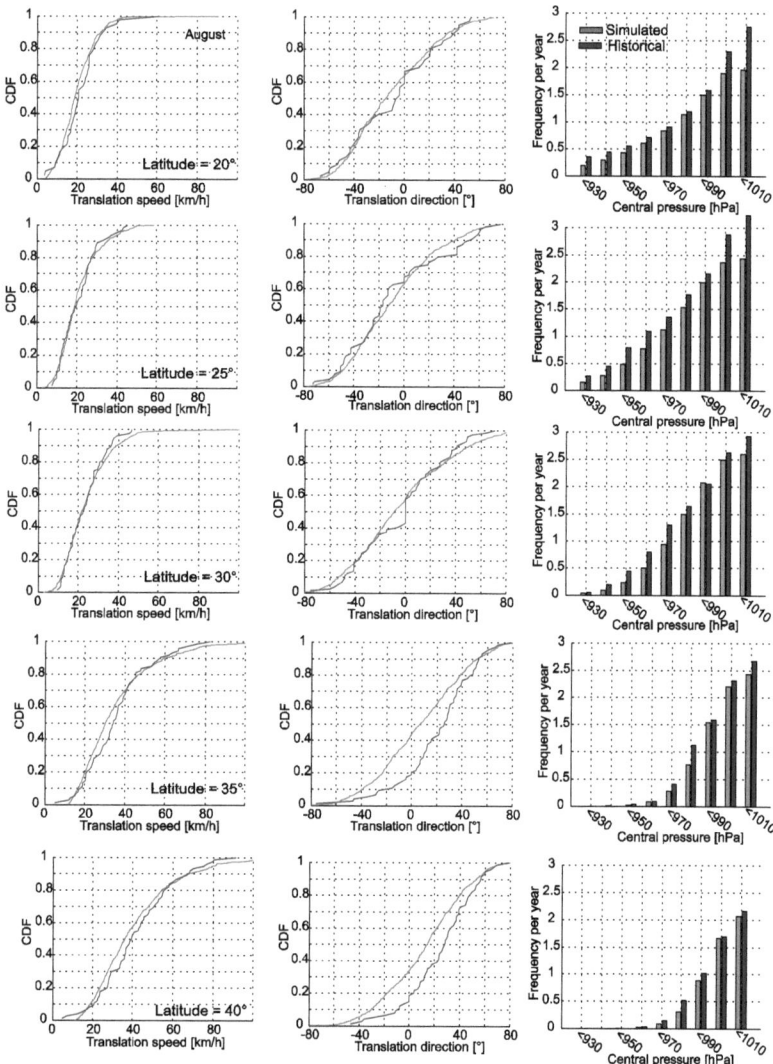

Figure 10.2: Cumulative distributions of the translation speed and direction and central pressure of typhoons in August.

10. Appendix

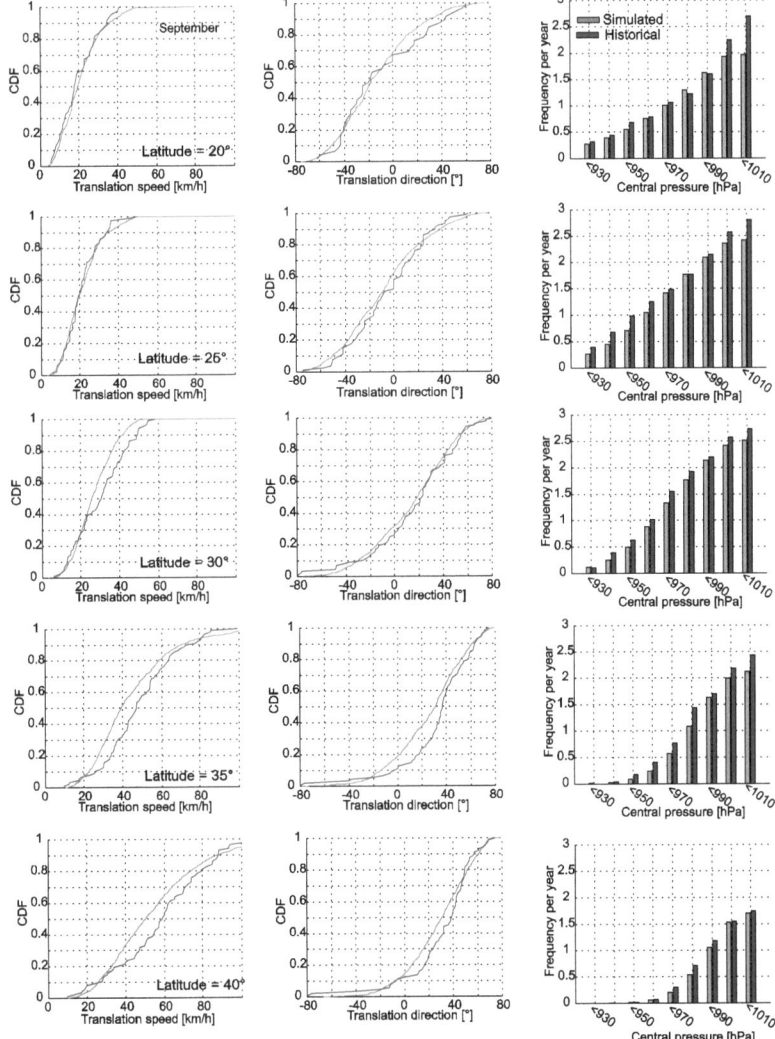

Figure 10.3: Cumulative distributions of the translation speed and direction and central pressure of typhoons in September.

10.2. Appendix B - Parameters for estimating the topography factor

Table 10-1: Parameters determining Eg (escarpments)

θ_S		X_S/H_S									
		-4	-2	-1	-0.5	0	0.5	1	2	4	8
7.5°	C_1	1.15	1.3	1.5	1.5	1.6	1.45	1.3	1.3	1.2	1.15
	C_2	0.8	0.8	0.8	0.8	0.8	0.7	0.6	0.6	0.5	0.4
	C_3	-2	-2	-2	-2	-2	-2	-2	-2	-2	-2
15°	C_1	0.4	1	1.2	1.55	2.1	1.65	1.5	1.3	1.2	1.15
	C_2	0.9	0	0.65	0.85	1	0.8	0.7	0.55	0.45	0.35
	C_3	-2	-2	-2	-2	-2	-2	-2	-2	-2	-2
30°	C_1	0.7	-0.5	1.05	1.1	1.3	1.3	1.25	1.2	1.115	1.1
	C_2	0.65	1.2	1.65	1.5	1.45	1.3	0.9	0.9	0.85	0.6
	C_3	-2	-2	1	0.8	0.3	0.3	0.5	0.7	1.2	1.4
45°	C_1	0.8	0	-3.5	1.1	1.2	1.35	1.3	1.2	1.15	1.1
	C_2	0.5	1	1.6	2	1.1	1.3	1.3	1.3	0.9	0.55
	C_3	-2	-2	-2	0.8	0.3	0.2	0.75	1.05	1.4	2
60°	C_1	0.6	0.1	-1.8	-2.4	1.2	1.4	1.35	1.25	1.15	1.1
	C_2	0.65	0.9	1.3	2.6	2	1.8	1.7	1.5	0.85	0.45
	C_3	-2	-2	-2	-1	0.5	0.5	0.8	1.2	1.9	3.1

Table 10-2: Parameters determining Eg (ridge-shaped topography)

θ_S		X_S/H_S									
		-4	-2	-1	-0.5	0	0.5	1	2	4	8
7.5°	C_1	1.1	1.2	1.35	1.35	1.4	1.3	1.3	1.2	1.1	1
	C_2	1	1	1	1	1.5	1.2	1.1	2	1.6	0
	C_3	0	0	0	0	0.2	0.2	0.2	0.5	0.9	0
15°	C_1	1	1.05	1.2	1.25	1.3	1.4	1.3	1.25	0.35	0.65
	C_2	0	0	1	1	1	1.5	1.5	2	3	2
	C_3	0	0	0	0	0	0.5	0.6	1.1	0.2	0.3
30°	C_1	0.75	0.55	0.85	1	1.2	1.3	1.25	1.2	1.1	1.02
	C_2	1.5	2	2	0	1	2	2	1.6	1.7	1.7
	C_3	0	0	0	0	0	1.1	1.3	2.1	2.2	2.8
45°	C_1	0.75	0.55	0.2	0.75	1.15	1.2	1.15	1.12	1.11	1.02
	C_2	1.5	2	2	3	1	2.5	2.5	2	1.6	1.3
	C_3	0	0	0	0	0	1.2	1.9	2.2	2.5	3.2
60°	C_1	0.75	0.55	0.2	0.2	1.15	1.12	1.15	1.12	1.1	1.02
	C_2	1.5	1.5	1.8	3	1	2.2	2.5	2	1.6	1.3
	C_3	0	0	0	0	0	1.8	2	2.3	2.6	3.4

10.3. Appendix C - Equivalence of these Bayesian probabilistic networks in Section 5.5

The equivalence between the Bayesian probabilistic networks in Figure 7 (left) and (right) can be shown in the following. Since a transformation converts the lognormal distribution into a normal distribution it is sufficient to show the equivalence considering normal distributed random variables. Assume two probabilistic representations:

Representation 1 :

$$\begin{cases} X = \mu_X + E \\ E \sim N(\mu_E, \sigma_E) \\ Y_i \sim N(X, \sigma_X), \quad (i=1,2,..,n) \end{cases} \quad (9.1)$$

Representation 2:

$$\begin{cases} Y_i = X_i + E, \quad (i=1,2,..,n) \\ E \sim N(\mu_E, \sigma_E) \\ X_i \sim N(\mu_X, \sigma_X), \quad (i=1,2,..,n) \end{cases} \quad (9.2)$$

where $E \sim N(\mu_E, \sigma_E)$ represents that the random variable E follows a normal distribution with the mean μ_E and the standard deviation σ_E. $X_i \sim N(\mu_X, \sigma_X)$ and $Y_i \sim N(X, \sigma_X)$, $(i=1,2,..,n)$ should be interpreted in the same manner. It is assumed that the Y_i's in Equation (9.1) are independent

10. Appendix

and that the X_i's in Equation (9.2) are independent respectively. To show the equivalence it is then sufficient to show that the joint probability density functions of E and Y_i, $(i=1,2,...,n)$ are identical.

The joint probability density function in Representation 1 is calculated as:

$$\begin{aligned}
f(y_1, y_2,..., y_n, e) &= \int f(y_1, y_2,..., y_n, e, x)\delta(x-e-\mu_X)dx \\
&= \int \prod_{i=1}^{n} f(y_i \mid e, x)\delta(x-e-\mu_X)f(e)dx \\
&= \int \prod_{i=1}^{n} \phi(y_i; x, \sigma_X)\delta(x-e-\mu_X)f(e)dx \\
&= \prod_{i=1}^{n} \phi(y_i; \mu_X + e, \sigma_X)\phi(e; \mu_E, \sigma_E)
\end{aligned} \qquad (9.3)$$

where $f(\cdot)$ and $f(\cdot \mid \cdot)$ represent the unconditional and conditional joint probability density functions respectively for the variables indicated in the augments, $\delta(\cdot)$ is the Dirac delta function and $\phi(\cdot; \mu_Z, \sigma_Z)$ is the probability density function of the normal distributed random variable with mean μ_Z and standard deviation σ_Z. The joint probability density function in Representation 2 is calculated as:

10. Appendix

$$f(y_1, y_2, ..., y_n, e) = \int \prod_{i=1}^{n} f(y_i | x_i, e) f(x_i, e) dx_1 ... dx_n$$

$$= \int \prod_{i=1}^{n} \delta(y_i - x_i - e) f(x_i) f(e) dx_1 ... dx_n$$

$$= \int \prod_{i=1}^{n} \delta(y_i - x_i - e) \phi(x_i; \mu_X, \sigma_X) f(e) dx_1 ... dx_n \quad (9.4)$$

$$= \prod_{i=1}^{n} \phi(y_i - e; \mu_X, \sigma_X) f(e)$$

$$= \prod_{i=1}^{n} \phi(y_i; \mu_X + e, \sigma_X) \phi(e; \mu_E, \sigma_E)$$

Therefore the two probabilistic representations are equivalent.

10.4. Appendix D - *TRAST* visualizations

Figure 10.4 shows several examples of the visualization of the typhoon track and the corresponding wind field generated by the software tool *TRAST*.

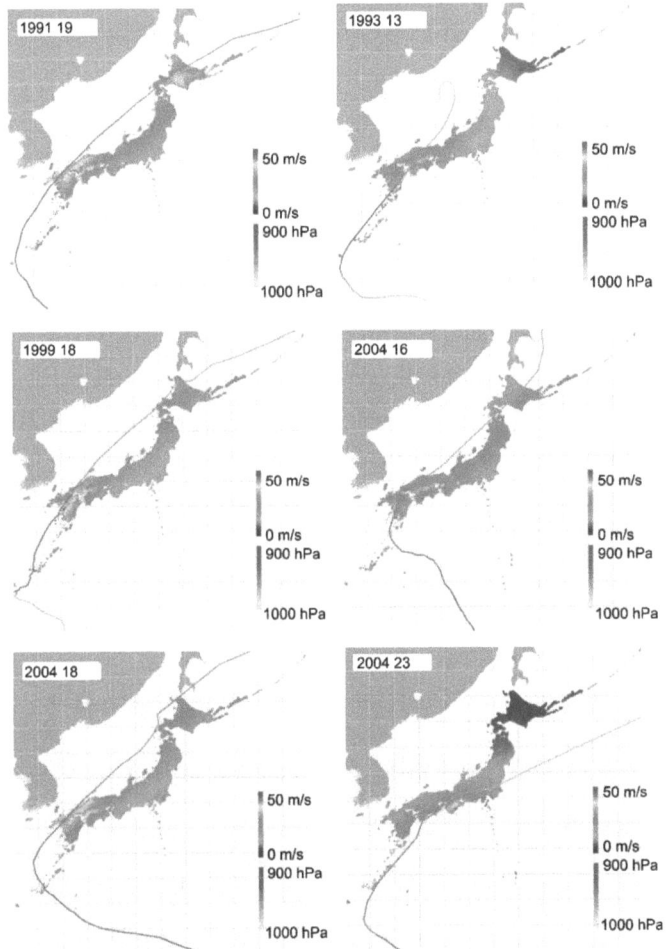

Figure 10.4: Visualization of historical typhoon tracks and wind fields

11. Nomenclature

Typhoon model

V_i is the typhoon translation speed [km/h] at time step.

Γ_i is the translation direction [°] of the typhoon.

$P_{C,i}$ is the central pressure of the typhoon at time step i.

T_i is the SST at the location where the considered typhoon is located at time step i.

ΔT_i is the pressure difference $T_{i+1} - T_i$.

a, b and c are the coefficients of the regression models.

$\varepsilon_V, \varepsilon_\Gamma$ and ε_{P_c} are the residual terms of the regression models.

ΔP_0 is the difference of the central pressure of a typhoon at the moment of the landfall and the peripheral pressure (here, 1013 [hPa] is assumed).

ΔP_t is the difference of the central pressure of the typhoon at time t and the peripheral pressure (here, 1013 [hPa] is assumed).

t is the time [hour] elapsed since the landfall.

d are the coefficients of the filling model.

R_M is radius of the maximum wind speed.

r is the distance from the center of the typhoon to the considered location.

α is the angle relative to the translation direction of the typhoon.

P_r is the pressure at the considered location r.

$\tilde{u}_g(r, \alpha)$ is the wind speed at gradient height at the location defined with r and α.

f is the Coriolis parameter assumed as $f = 1.46 \times 10^{-4} \times \sin\phi$ [1/s], where ϕ is the latitude of the representative location of the typhoon.

11. Nomenclature

ρ	is the air density assumed as $\rho = 1.275$ [kg/m³].
$u(z)$	is the *one-hour* sustained wind speed at the height of z.
z_g	is the gradient height.
$U_{s,10}$	is the 10-minute sustained wind speed at surface.
E_g	is the topological factor.
Ro_λ	is the modified Rossby number.
$\gamma(z)$	is the inflow angle at the height z.
G	is the gust factor.
u_{gust}	is the gust wind speed.
z_S	is the height of the measurement device [m].
t	is the duration of the record [sec.]
θ_s	is the inclination of the topographic feature.
$X_s(m)$	is the distance of the topographic feature.
$H_S(m)$	is the height of the topographic feature.
$Z(m)$	is the height above ground of the considered object.

Treatment of epistemic uncertainties in the typhoon model

Θ	are the epistemic random variables.
$\mathbf{X}$	are the aleatory random variables.

Updating of the typhoon model

H	represents the occurrence of a hazard event.
F_i	represents the state of a structure (*failure* or *no failure*, for instance) at the ith location during a typhoon event.
D	is the random variable that represents the parameters of the fragility model.
W_i	represents the maximum wind speed at the ith location during a typhoon event.

11. Nomenclature

p_i represents the probability of failure of the structure at the ith location.

M is the number of samples.

ε^j represents the model uncertainty for the jth typhoon event.

v_k^j,   are the measured maximum wind speeds at the meteorological stations k during the jth typhoon event.

Application: Portfolio risk analysis

Q is the ground-up loss ratio.

w is the hazard index (maximum wind speed).

s is the type of structure.

c is the cause of damages.

c_{pol} is the policy condition.

$^n V_{insured}$ is the insured value of the nth exposure.

$\varepsilon_{s,c}$ represents the model uncertainties explained.

$V_{franchise}$ is the franchise value.

Application: Global warming risk assessment

L is the probability distribution of the annual maximum wind load.

R is the probability distribution of the resistance.

V is the probability distribution of the annual maximum wind speed.

e_{SST} represents a scenario of SST increase.

p_F is the probability of failure.

$y_{NHM,MAX}$ is the maximum wind speed at 10 [m] height from the surface reproduced by the JMA-NHM.

$r_d(y_{NHM,MAX})$ is the damage ratio as a function of $y_{NHM,MAX}$.

N_j^B is the number of buildings at location j.

11. Nomenclature

N_j is the number of buildings that are damaged by typhoons in a given year at location j.

$\bar{R}_j$ is the building risk at location j.

Application: Risk assessment of a approaching typhoon and real-time decision making

X represents the hazard index.

Y_i represents the variables characterizing the phenomena.

E_i are the information.

U is the utility function.

A_i are the alternative decision.

C_D are the damage costs.

C_{PI} are the interruption costs.

12. References

Anders, A. and K. Nishijima (2011) *Adaption of option pricing algorithm to real time decision optimization in the face of emerging natural hazards* 11th International Conference on Applications of Statistics and Probability in Civil Engineering, Zürich, Switzerland.

Ando, T. (2010). *Bayesian model selection and statistical modeling*, Chapman & Hall.

Ando, T. and R. Tsay (2009) *Predictive likelihood for Bayesian model selection and averaging*, International Journal of Forecasting, 26 (4), pp. 744-763.

Architectural Institute of Japan (2004). *AIJ Recommendations for Loads on Buildings*.

AS1170.2 (1989) *Standard.*, Australia.

ASCE7-98 (2000) *Minimum design loads for buildings and other structures,*, Revision of ANSI/ASCE.

Ashcroft, J. (1994) *The relationship between the gust ratio, terrain roughness, gust duration and the hourly mean wind speed*, Journal of Wind Engineering and Industrial Aerodynamics, 53 (3), pp. 331-355.

12. References

Atallah, E. H. and L. F. Bosart (2003) *The Extratropical Transition and Precipitation Distribution of Hurricane Floyd (1999)*, Monthly Weather Review, 131 (6), pp. 1063-1081.

Bartheelmie, R. J., J. P. Palutikof and T. D. Davies (1993) *Estimation of sector roughness lengths and the effect on prediction of the vertical wind speed profile* Boundary-Layer Meteorology, 66 (1-2), pp. 19-47.

Basöz, N. I., A. S. Kiremidjian, S. A. King and K. H. Law (1999) *Statistical Analysis of Bridge Damage Data from the 1994 Northridge*, Earthquake Spectra, 15 (1), pp. 25-54.

Batt, M. E., et al., (1980) *Hurricane wind speeds in the United States.*, National Bureau of Standards, US Departoment of Commerce.

Brand, S., K. Rabe and T. Laevastu (1977) *Parameterization Characteristics of a Wind-Wave Tropical Cyclone Model for the Western North Pacific Ocean*, Journal of Physical Oceanography, 7 (5), pp. 739-746.

Bretschneider, C. L. (1959) *Hurricane design-wave practices*, Trans. ASCE, 124, pp. 39-62.

Brooks, S. P. (1998) *Markov Chain Monte Carlo Method and Its Application*, The Statistician, 47 (1), pp. 69-100.

12. References

Burri, D., C. Celio, M. Graf and M. H. Faber (2009) *Neuberechnung der japanischen Windlast Norm*, Institut für Baustatik und Konstruktion, ETH Zürich.

Chen, W. (2003). *Earthquake engineering handbook*, CRC Press.

Congdon, P. (2006). *Bayesian Statistical Modelling*, Wiley series in probability and statistics.

Davenport, A. G. (1961) *The spectrum of horizontal gustiness near the ground in high winds*, Quarterly Journal of the Royal Meteorological Society, 87 (372), pp. 194-211.

Davenport, A. G. (1965) *The relationship of wind structure to wind loading*, Proc. Symp. No. 16, on Wind effects on building and structures, Nat. Phys. Lab.,.

De Groot, H. (1970). *Optimal Statistical Decisions*, John Wiley&Sons, Inc.

De Sanctis, G., M. Graf and M. H. Faber (2008) *Taifunmodellierung in Japan: Einflüsse topographischer Effekte*, Institut für Baustatik und Konstruktion, ETH Zürich.

DeMaria, M. (2009) *A Simplified Dynamical System for Tropical Cyclone Intensity Prediction*, Monthly Weather Review, 137 (1), pp. 68-82.

12. References

DeMaria, M. and J. Kaplan (1994) *A Statistical Hurricane Intensity Prediction Scheme (SHIPS) for the Atlantic Basin*, Weather and Forecasting, 9.

DeMaria, M. and J. Kaplan (1999) *An Updated Statistical Hurricane Intensity Prediction Scheme (SHIPS) for the Atlantic and Eastern North Pacific Basins*, Weather and Forecasting, 14 (3), pp. 326-337.

DeMaria, M., M. Mainelli, L. K. Shay, J. A. Knaff and J. Kaplan (2005) *Further Improvements to the Statistical Hurricane Intensity Prediction Scheme (SHIPS)*, Weather and Forecasting, 20 (4), pp. 531-543.

Department of Homeland Security (2011) *HAZUS MHMR5 Technical Manual*.

Eklund, J. and S. Karlsson (2004) *Forecast Combination and Model Averaging Using Predictive Measures*

Emanuel, K. (2006a) *A Statistical Deterministic Approach to Hurricane Risk Assessment* American Meteorological Society, pp. 299-314.

Emanuel, K. (2006b) *A Statistical Deterministic Approach to Hurricane Risk Assessment Supplement*, American Meteorological Society, pp. 299-314.

Emanuel, K. A. (1987) *The dependence of hurricane intensity on climate*, Nature, 326 (6112), pp. 483–485.

Emanuel, K. A. (1988) *The Coupled Hurricane Intensity Prediction Scheme (CHIPS)*.

Emanuel, K. A. (2005) *Increasing destructiveness of tropical cyclones over the past 30 years*, Nature, 436 (3906), pp. 686-688.

Faber, M. H. (2005) *On the Treatment of Uncertainties and Probabilities in Engineering Decision Analysis*, Journal of Offshore Mechanics and Arctic Engineering, Trans. ASME, 127(3), pp. 243-248.

Faber, M. H., Y. Bayraktarli and K. Nishijima (2007) *Recent Developments in the Management of Risks Due to Large Scale Natural Hazards*, XVI Congreso Nacional Ingenieria Sismica, Ixtapa-Zihuatanejo, Mexico.

Fitzpatrick, P. j. (1997) *Understanding and Forecasting Tropical Cyclone Intensity Change with the Typhoon Intensity Prediction Scheme (TIPS)*, Weather and Forecasting, 12.

Frank, D. M. J. (1985). *Evolution of the structure of precipitation in Hurricane Allen (1980)*. Boston, MA, ETATS-UNIS, American Meteorological Society.

Friedman, D. G. (1975). *Computer Simulation of Natural Hazard Assessment*.

Fujii, T. (1998) *Statistical Analysis of the Characteristics of Severe Typhoons Hitting the Japanese Main Islands*, Monthly Weather Review, 126 (4), pp. 1091-1097.

Fujii, T. and Y. Mitsuta (1986) *Simulation of winds in typhoons by a stochastic model.*, Journal of Wind Engineering,, 28, pp. 1-12.

Gelman, A., J. B. Carlin, H. S. Stern and D. B. Rubin (2004). *Bayesian Data Analysis*, London: Chapman and Hall.

Gelman, A. and J. Hill (2007). *Data Analysis Using Regression and Multilevel/Hierarchical Models*, Cambridge.

Genie (2011) *Genie 2.0, http://genie.sis.pitt.edu/*.

Georgiou, P. N., A. G. Davenport and B. J. Vickery (1983) *Design wind speeds in regions dominated by tropical cyclones*, Journal of Wind Engineering and Industrial Aerodynamics, 13 (1-3), pp. 139-152.

Graf, M. and K. Nishijima (2011) *Issues of epistemic uncertainty treatment in decision analysis for tropical cyclone risk management*, 11th International Conference on Applications of Statistics and Probability in Civil Engineering, ICASP 11, Zürich, Switzerland.

Graf, M., K. Nishijima and M. H. Faber (2008) *Adaption of Typhoon Risk Modeling to Climate Changes*, IDRC 2008, Davos.

Graf, M., K. Nishijima and M. H. Faber (2009) *A Probabilistic Typhoon Model for the Northwest Pacific Region*, APCWE 7, Taipei.

Grimmond, C. S. B. and T. R. Oke (1999) *Aerodynamic Properties of Urban Areas Derived from Analysis of Surface Form*, Journal of Applied Meteorology, 38 (9), pp. 1262-1292.

Gyftodimos, E. and P. A. Flach (2002) *Hierarchical Bayesian Networks: A Probabilistic Reasoning Model for Structured Domains*, ICML-2002 Workshop on Development of Representations.

Hall, T. M. and S. Jewson (2007) *Statistical modelling of North Atlantic tropical cyclone tracks*, Tellus A, 59 (4), pp. 486-498.

Hamid, S., B. M. Golam Kibria, S. Gulati, M. Powell, B. Annane, S. Cocke, J.-P. Pinelli, K. Gurley and S.-C. Chen (2010) *Predicting losses of residential structures in the state of Florida by the public hurricane loss evaluation model*, Statistical Methodology, 7 (5), pp. 552-573.

Helliwell, N. C. (1971) *Wind over London*, Wind effects on buildings and structures, Tokyo, 1971.

Hoeting, J. A., D. Madigan, A. E. Raftery and C. T. Volinsky (1999) *Bayesian Model Averaging: A Tutorial*, Statistical Science, 14 (4), pp. 382-401.

Holland, G. J. (1980) *An analytical model of the wind and pressure profiles in hurricanes*, Mon. Weather Rev., 108, pp. 1212-1218.

12. References

Hugin. (2006). "*Hugin Expert A/S Version 6.6 Software, www.hugin.com.*"

Ishihara, T., K. K. Siang, C. C. Leong and Y. Fujino (2005) *Wind Field Model and Mixed Probability Distribution Function for Typhoon Simulation*, The Sixth Asia-Pacific Conference on Wind Engineering (APCWE-VI), Seoul, Korea.

Itoi, T., J. Kanda and H. Choi (2005) *Estimation of vertical mean wind speed profile from statistical characteristics of obstacles on ground surface.*, Journal of Structural and Construction Engineering, 587, pp. 45-52.

IWTC-VII (2010) *7th International Workshop on Tropical Cyclones (IWTC-VII) report.*

JCSS. (2001). "*Probabilistic Model Code, The Joint Committee on Structural Safety.*" Retrieved 2008, 2001, from http://www.jcss.ethz.ch.

Jensen, F. V. (2001). *Bayesian Networks and Decision Graphs*, New York: Springer.

Jensen, F. V. and T. D. Nielsen (2007). *Bayesian Networks and Decision Graphs*, New York: Springer.

Katsuchi, H. and H. Yamada (2005) *Typhoon Simulation Technique incorporating Sea-surface Temperature*, 10th American Conference on Wind Engineering, Baton Rouge.

Kjaerulff, U. (1995) *dHugin: A computational system for dynamic time-sliced Bayesian networks*, Inteternaltional Journal of Forecasting.

Klugman, S. A., H. H. Panjer and G. E. Willmot (2004). *Loss Models from data to desicions*, Wiley.

Knaff, J. A., C. R. Sampson and M. DeMaria (2005) *An Operational Statistical Typhoon Intensity Prediction Scheme for the Western North Pacific*, Weather and Forecasting, 20 (4), pp. 688-699.

Knutson, T. R., J. L. McBride, J. Chan, K. Emanuel, G. Holland, C. Landsea, I. Held, J. P. Kossin, A. K. Srivastava and M. Sugi (2010) *Tropical cyclones and climate change*, Nature Geosci, 3 (3), pp. 157-163.

Knutson, T. R. and R. E. Tuleya (2004) *Impact of CO_2-induced warming on simulated hurricane intensity and precipitation: Sensitivity to the choice of climate*, Journal of Climate, 17 (18), pp. 3477-3490.

Kondo, J. and H. Yamazawa (1986) *Aerodynamic roughness over an inhomogeneous ground surface* Boundary-Layer Meteorology, 35 (4).

Kwok, K. C. S. and P. A. Hitchcock (2009) *Characterisation of and wind-induced pressures in a compartmentalised building during a typhoon*, Journal of Wind and Engineering, 6 (2).

Lee, K. H. and D. V. Rosowsky (2007) *Synthetic Hurricane Wind Speed Records: Development of a Database for Hazard Analyses and Risk Studies*, Natural Hazards Review, 8 (2), pp. 23-34.

Lettau, H. (1969) *Note on Aerodynamic Roughness-Parameter Estimation on the Basis of Roughness-Element Description*, Journal of Applied Meteorology, 8 (5).

Lettau, H. (1970) *Physical and meteorological basis for mathematical models of urban diffusion processes*, Proc. Symp. on Multiple-source urban diffusion models, U.S. EPA, Publ. AP-86 Research Triangle Park, N.C., 1970.

Lunn, D. J., A. Thomas, N. Best and Spiegelhalter (2000) *WinBUGS - A Bayesian modelling framework: concepts, structure, and extensibility*, Statistics and Computing, 10, pp. 325-337.

Maruyama, T., E. Tomokiyo and J. Maeda (2010) *Simulation of Strong Wind Field by Non-hydrostatic Mesoscale Model and Its Applicability for Wind Hazard Assessment of Buildings and Houses*, Hydrological Research Letters, 4, pp. 40-44.

Matsui, M., Y. Meng and K. Hibi (1998) *Extreme typhoon wind speeds considering differences in the average time between full-scale observations and typhoon model*, Journal of Structural and Construction Engineering, 506, pp. 67-74.

Matsui, M., Y. Tamura and S. Tanaka (2002) *Wind load of tall buildings considering wind directionality effects.*, National Symposium on Wind Engineering, Japan, 17th, pp. 499-504.

Melchers, R. E. (2001). *Structural reliability analysis and prediction*, University of Newcastle, Australia, Wiley.

Meng, Y., M. Matsui and K. Hibi (1995a) *Analytical model for simulation of the wind field in a typhoon boundary layer*, Journal of Wind Engineering and Industrial Aerodynamics, 56, pp. 291-310.

Meng, Y., M. Matsui and K. Hibi (1995b) *An analytical model for simulation of the wind field in a typhoon boundary layer*, Journal of Wind Engineering and Industrial Aerodynamics, 56, pp. 291-310.

Meng, Y., M. Matsui and K. Hibi (1997) *A numerical study of the wind field in a typhoon boundary layer*, Journal of Wind Engineering and Industrial Aerodynamics, 67&68, pp. 437-448.

Miller, B. I. (1967) *Characteristics of Hurricanes*, Science, 157 (3795), pp. 1389-1399.

Mizuta, R., H. Yoshimura, H. Murakami, M. Matsueda, H. Endo, T. Ose, K. Kamiguchi, M. Hosaka, M. Sugi, S. Yukimoto, S. Kusunoki and A.

Kitoh (2011) *Climate simulations using MRI-AGCM3.2 with 20-km grid*, Journal of the Meteorological Society of Japan, pp. (Submitted).

Murakami, H. and M. Sugi (2010) *Effect of Model Resolution on Tropical Cyclone Climate Projections*, Scientific Online Letters on the Atmosphere, 6, pp. 73-76.

Ngo, T. and C. Letchford (2008) *A comparison of topographic effects on gust wind speed*, Journal of wind engineering and industrial aerodynamics.

Nishijima, K. and M. H. Faber (2007) *A Bayesian framework for typhoon risk management*, 12th International Conference on Wind Engineering, Cairns, Australia.

Nishijima, K., M. Graf and M. H. Faber (2008a) *From Near-real-time Information Processing to Near-real-time Decision Making in Risk Management of Natural Hazards*, Inaugural International Conference of the Engineering Mechanics Institute, EM08, University of Minnesota, Minneapolis, Minnesota.

Nishijima, K., M. Graf and M. H. Faber (2009) *Optimal evacuation and shut-down decisions in the face of emerging natural hazards*, ICOSSAR 2009, Osaka, Japan.

Nishijima, K., T. Itoi and J. Kanda (2004) *Modeling of strong wind speed driven by typhoon and its spatial dependency with multivariate extreme value distribution.*, Cherry Bud workshop, Yokohama, Japan.

12. References

Nishijima, K., M. Maes and M. H. Faber (2008b) *Probabilistic assessment of extreme events subject to epistemic uncertainties*, ASME 27th International Conference on Offshore Mechanics and Arctic Engineering, OMAE2008, Estoril, Portugal.

Nishijima, K., T. Maruyama and M. Graf (2011) *A preliminary impact assessment of typhoon wind risk of residential buildings in Japan under future climate change*, submitted to Hydrological Research Letters.

O'Rourke, M. J., M. Eeri and P. So (2000) *Seismic Fragility Curves for On-Grade Steel Tanks*, Earthquake Spectra, 16 (4), pp. 801-815.

Paté-Cornell, M. E. (1996) *Uncertainties in risk analysis: Six levels of treatment*, Reliability Engineering & System Safety, 54 (2-3), pp. 95-111.

Pinelli, J., E. Simiu, K. Gurley, C. Subramanian, L. Zhang, A. Cope, J. J. Filliben and S. Hamid (2004) *Hurricane Damage Prediction Model for Residential Structures* Journal of Structural Engineering 130 (11).

Raftery, A. E., D. Madigan and J. A. Hoeting (1997) *Bayesian Model Averaging for Linear Regression Models*, Journal of the American Statistical Association, 92 (437), pp. 179-191.

Raiffa, H. and R. Schlaifer (1961). *Applied Statistical Decision Theory*. Cambridge, Cambridge University Press.

Riggelsen, C., N. Gianniotis and F. Scherbaum (2011) *Learning aggregations of ground-motion models using data*, ICASP 11, Zürich, Switzerland.

Ross, D. (1976) *A simplified model for forecasting hurricane generated waves.*, Bull. Am. Meteorolog. Soc., 57 (1).

Rossetto, T. and A. Elnashai (2003) *Derivation of vulnerability functions for European-type RC structures based on observational data*, Engineering Structures, 25, pp. 23.

Rumpf, J., H. Weindl, P. Höppe, E. Rauch and V. Schmidt (2007) *Stochastic modelling of tropical cyclone tracks*, Mathematical Methods of Operations Research, 66 (3), pp. 475-490.

Russell, L. R. (1971) *Probability distributions for hurricane effects*, J. of the Waterways, Harbor and Coastal Engineering Division, 97 (1), pp. 139-154.

Schloemer, R. W. (1954a) *Analysis and synthesis of hurricane wind patterns over Lake Okeechobee, Florida*, Hydrometeorological Report, 31, pp. 49.

Schloemer, R. W. (1954b) *Analysis and synthesis of hurricane wind patterns over Lake Okeechobee, Florida*, Hydrometeorological Report, Vol. 31, pp. 49.

12. References

Shapiro, L. J. (1983) *The Asymmetric Boundary Layer Flow Under a Translating Hurricane*, Journal of the Atmospheric Sciences, 40, pp. 1984-1998.

Shinozuka, M., M. Q. Feng, J. Lee and T. Naganuma (2000) *Statistical analysis of fragility curve*, Journal of Engineering Mechanics, Trans ASCE 2000, 126 (12), pp. 1224-1231.

Simiu, E., J. F. Nash and V. C. Patel (1976) *Mean Speed Profiles of Hurricane Winds* Journal of the Engineering Mechanics Division, 102 (2).

Simiu, E. and R. H. Scanlan (1996) *Wind effects on structures: Fundamental and applications to design*, New York: Wiley.

Solomon, S., D. Qin, M. Manning, R. B. Alley, T. Berntsen, N. L. Bindoff, Z. Chen, A. Chidthaisong, J. M. Gregory, G. C. Hegerl, M. Heimann, B. Hewitson, B. J. Hoskins, F. Joos, J. Jouzel, V. Kattsov, U. Lohmann, T. Matsuno, M. Molina, N. Nicholls, J. Overpeck, G. Raga, V. Ramaswamy, J. Ren, M. Rusticucci, R. Somerville, T. F. Stocker, P. Whetton, R. A. Wood and D. Wratt (2007) *Technical Summary. In: Climate Change 2007: The Physical Science Basis. Contribution of Working Group I to the Fourth Assessment Report of the Intergovernmental Panel on Climate Change*, Intergovernmental Panel on Climate Change, pp. 73-74.

Straub, D. and A. Der Kiureghian (2007) *Improved seismic fragility modeling from empirical data*, Structural Safety, 30 (4), pp. 320-336.

12. References

Tomokiyo, E. and J. Maeda (2004) *CFD prediction of local winds associated with Typhoon Tokage (2004) Effects of atmospheric stability on the increase of strong winds*, The Fifth International Symposium on Computational Wind Engineering (CWE2010), Chapel Hill, North Carolina, USA.

Tomokiyo, E., J. Maeda and N. Tsuru (2009) *Wind Disaster in Kyushu due to Typhoons in 2004 - Residential Damage in Kyushu, Japan*, APCWE-VII, Taipei, Taiwan.

Tryggvason, B. V., A. G. Davenport and D. Surry (1976) *Predicting wind-induced response in hurricane zones*, Proc. ASCE, Jour. Struc. Div., 102 (12), pp. 2333-2350.

Verkaik, J. W. (2000) *Evaluation of two gustiness models for exposure correction calculations*, American Meteorological Society.

Vickery, P. J. (2005) *Simple Empirical Models for Estimating the Increase in the Central Pressure of Tropical Cyclones after Landfall along the Coastline of the United States*, Journal of Applied Meteorology, 44 (12), pp. 1807-1826.

Vickery, P. J., J. Lin, P. F. Skerlj, L. A. Twisdale and K. Huang (2006a) *HAZUS-MH Hurricane Model Methodology. I: Hurricane Hazard, Terrain, and Wind Load Modeling* NATURAL HAZARDS, (82).

Vickery, P. J. and P. F. Skerlj (2005) *Hurricane Gust Factors Revisited*, Journal of Structural Engineering, 131 (5), pp. 825-832.

Vickery, P. J., P. F. Skerlj, J. Lin, L. A. Twisdale, M. A. Young and F. M. Lavelle (2006b) *HAZUS-MH Hurricane Model Methodology. II: Damage and Loss Estimation.*, Natural Hazards Review,, 7 (94).

Vickery, P. J., P. F. Skerlj and L. A. Twisdale (2000) *Simulation of Hurricane Risk in the U.S. Using Empirical Track Model*, Journal of Structural Engineering, 126 (10), pp. 1222-1237.

Vickery, P. J. and L. A. Twisdale (1995a) *Prediction of Hurricane Wind Speeds in the United States* Journal of Structural Engineering 121 (11).

Vickery, P. J. and L. A. Twisdale (1995b) *Wind-Field and Filling Models for Hurricane Wind-Speed Predictions*, Journal of Structural Engineering, Vol. 121 (11), pp. 1700-1709.

Vickery, P. J. and L. A. Twisdale (1995c) *Wind-Field and Filling Models for Hurricane Wind-Speed Predictions*, Journal of Structural Engineering, 121 (11), pp. 1700-1709.

Walker, G. R. (1997) *Current developments in catastrophe modeling*, Financial Risk Management for Natural Catastrophes, eds. N.R. Britton and J. Oliver, Aon Group Australia Limited, Griffith University, Brisbane (1997), pp. 17-35.

12. References

Watson, C. C. and M. E. Johnson (2004). *Hurricane loss estimation models: Opportunities for improving the state of the art*. Boston, MA, US, American Meteorological Society.

Wieringa, J. (1976) *An objective exposure correction method for average wind speeds measured at a sheltered location*, Quarterly Journal of the Royal Meteorological Society, 102 (431), pp. 241-253.

Wieringa, J. (1986) *Roughness-dependent geographical interpolation of surface wind speed averages*, Quarterly Journal of the Royal Meteorological Society, 112 (473), pp. 867-889.

Wieringa, J. (1993) *Representative roughness parameters for homogeneous terrain* Boundary-Layer Meteorology, 63 (4).

World Meteorological Organization (2006) *The status of the global climate in 2006*.

Yasui, H., T. Ohkuma, H. Marukawa and J. Katagiri (2002) *Study on evaluation time in typhoon simulation based on Monte Carlo method*, Journal of Wind Engineering and Industrial Aerodynamics, 90 (12–15), pp. 1529-1540.

Yin, J., M. B. Welch, H. Yashiro and M. Shinohara (2009) *Basinwide Typhoon Risk Modeling and Simulation for West North Pacific Basin*, APCWE 7, Taipei.

Yonekura, E. and T. M. Hall (2011) *A Statistical Model of Tropical Cyclone Tracks in the Western North Pacific with ENSO-Dependent Cyclogenesis*, Journal of Applied Meteorology and Climatology, 50 (8), pp. 1725-1739.

Yoshizumi, S. (1968) *On the Asymmetry of Wind Distribution in the Lower Layer in Typhoon.*, Journal of the Meteorological Society of Japan, 46 (3), pp. 153-159.

Young, I. R. (1988) *Parametric Hurricane Wave Prediction Model*, Journal of Waterway, Port, Coastal, and Ocean Engineering, 114 (5), pp. 637-652.

List of Figures

Figure 1.1: Integration of knowledge, data and information 17

Figure 2.1: Components of the developed typhoon model. 39

Figure 2.2: Geographical distribution of the location of the initiation of historical typhoon events. 46

Figure 2.3: Bayesian probabilistic network for the occurrence of typhoons.47

Figure 2.4: Grids and months for the probabilistic model for translation. 49

Figure 2.5: 18 zones for the probabilistic model for central pressure. (The model is developed for each individual month in the same way as the model for translation, but it is not illustrated in the figure.). 49

Figure 2.6: Number of data 51

Figure 2.7: Translation speed 52

Figure 2.8: Interpolation area 54

Figure 2.9: Coordinate system used in the wind field model. 58

Figure 2.10: Example of the wind field calculated using the wind field model. 59

Figure 2.11: Conversions of wind speeds ($\tilde{u}_g$ is the abbreviation of $\tilde{u}_g(r,\alpha)$ for any given location) 62

Figure 2.12: Coordinate system used in the wind field model. 62

Figure 2.13: Coordinate system in vertical direction. 65

Figure 2.14: Coordinate system used in the surface friction model. 66

Figure 2.15: Wind field at gradient height (left) and wind field at surface height (right) of the typhoon Songda 2004 18. 67

Figure 2.16: Land use data 69

Figure 2.17: Land use data and map of city centers 69

Figure 2.18:Considered segment for estimating the roughness length 70

Figure 2.19: Escarpments 73

Figure 2.20: Ridge-shaped topography 73

Figure 2.21: Topographic map 74

Figure 3.1: Comparison of the occurrence of typhoons 77

Figure 3.2: Lines and area which the probabilistic characteristics of typhoons travelling through are compared. 78

Figure 3.3: Cumulative mean frequencies of typhoons crossing different latitudes between longitudes $[120, 160°]$ for August and September. 79

Figure 3.4: Cumulative distributions of the translation speed and direction of typhoons crossing the latitude of $30°$ between the longitudes $[120, 160°]$ in September. 80

Figure 3.5: Mean frequency of the landfalls of typhoons in different months. 81

Figure 3.6: Verification of the procedure for the development of the occurrence and transition models. 82

Figure 3.7: Validation of the approach employed for the development of the occurrence and transition models. 82

Figure 3.8: Extrapolation of the typhoon model to the future 83

Figure 3.9: Comparison of the effect of considering the seasonal difference in the transition model for all months. 85

Figure 3.10: Comparison of the effect of considering the seasonal difference in the transition model for July. 85

Figure 3.11: Comparison of the effect of considering the seasonal difference in the transition model for August. 86

12. References

Figure 3.12: Comparison of the effect of considering the seasonal difference in the transition model for September. 86

Figure 3.13: Time histories of the wind speeds, wind directions and the maximum wind speeds at two meteorological stations during several historical typhoon events. 88

3.14: Time histories of the wind speeds, wind directions and the maximum wind speeds at two meteorological stations during several historical typhoon events. 89

Figure 3.15: Locations at which 100-year and 500-year wind speeds are compared in Table 3-1. 91

Figure 4.1: Probabilistic assessment subject to aleatory and epistemic uncertainties (after Nishijima et al. (2008b)). 99

Figure 4.2: Framework to integrate epistemic uncertainties due to model selection into the decision analysis. 100

Figure 4.3: Components of the developed typhoon model. 102

Figure 4.4: Spatial grids and temporal slices for the probabilistic model for translation. 105

Figure 4.5: Spatial zones and temporal slices for the probabilistic model for central pressure. 105

Figure 4.6: Line segments at which the numbers of typhoons are counted. 109

Figure 4.7: Expected numbers of typhoons that intersect the line segments at different latitudes. 109

Figure 4.8: Cumulative distributions of the direction of movement (left), of the translation speed (centre) and of the central pressure of typhoons (right) crossing the latitude of 30° between the longitude [120,160°] in September for the different alternative models. 110

Figure 4.9: Cumulative distributions of the direction of movement (left), of the translation speed (centre) and of the central pressure (right) of typhoons crossing the latitude of 35° between the longitude [120,160°] in September for the different alternative models. 110

Figure 4.10: Cumulative distributions of the direction of movement of typhoons crossing the latitude of 30° between the longitude [120,160°] in September for the alternative models with different discretization (left), different functional form (centre) and different data set (right). 111

Figure 4.11: Maximum wind speeds as a function of return period for Ishigaki (left), Tokyo (centre) and Sapporo (right) for the different alternative models. 111

Figure 4.12: Maximum wind speeds as a function of return period for Tokyo for the alternative models with different discretization (left), different functional form (centre) and different data set (right). Variation of the maximum wind speed of the alternative models . 112

Figure 4.13: Locations of the cities where the annual maximum 10-min sustained wind speed are compared. 113

Figure 5.1: Diagram of risk management of natural hazards. 120

Figure 5.2: An illustration of a Bayesian probabilistic network (left) with instantiated nodes (center) and an equivalent abbreviated representation of the network (right). 122

Figure 5.3: Considered Bayesian probabilistic network in Example 4.1 with instantiated nodes for the proposed approach (left) and for the standard and ideal approach (right). 125

Figure 5.4: Rate of convergence of the parameter a as a function of the sample size used for the updating. 130

Figure 5.5: Rate of convergence of the parameter b as a function of the sample size used for the updating. 130

Figure 5.6: Fragility curve estimated by the different approaches for different ξ_w. 131

Figure 5.7: Bayesian probabilistic network including a model uncertainty ε for the proposed approach used in Example 4.2 (left) and Bayesian probabilistic network used in Example 4.3 (right). 132

Figure 5.8: Mean value of the posterior distribution of parameter a after updating with datasets with different values of the parameter ρ (left) and with datasets with samples from different number of typhoon events (right). 132

Figure 5.9: Rate of convergence of the parameter a (left) and b (right) as a function of the sample size used for updating. 134

Figure 6.1: Flow of the development of the vulnerability model. 144

Figure 6.2: Loss ratio as a function of wind speed given the occurrence of loss (type of structure: Wood, types of object: building) 147

Figure 6.3: Ratios of the occurrence of loss and the estimated probabilities of the occurrence of loss. 151

Figure 6.4: Median-median ground-up loss ratios for exposures whose type of object is "building". 152

Figure 6.5: Comparison of historical losses and reproduced losses using the developed typhoon model. 154

Figure 6.6: Example of the relation between ground-up loss ratio and insured loss ratio for a given wind speed. 159

Figure 6.7: Diagram on the disaggregation of aggregated exposure data. 168

Figure 6.8: Example of the disaggregation in the area of Tokyo. 168

Figure 6.9: Components of programs. 170

Figure 6.10: Graphical user interface (Main tab window). 171

Figure 6.11: Insurance payment of the policy conditions described by the policy data file in the format "A". 173

Figure 6.12: Insurance payments of the two policy conditions described by the policy data file in the format "B". 174

Figure 6.13: Components concerning the analysis settings. 175

Figure 6.14: Summary of the analysis results for the case of the analysis with the stochastic event set. 176

Figure 6.15: Losses of individual typhoon events and visualization of track and wind speeds of a selected event. 177

Figure 6.16: Report generator. 180

Figure 6.17: Interface for the management of the analysis results. 181

Figure 7.1: Approach suggested to assess and mitigate the impact of extreme wind events caused by climatic change. 185

Figure 7.2: Estimation of the target probability of failure 189

Figure 7.3: Change of the characteristic value (98%-quantile value) of annual maximum wind speed (left) and the change of the probability of failure (right) 190

Figure 7.4: Adaption of structural design (left) and required change of the characteristic value (right). 190

Figure 7.5: Extracted typhoons for the current climate (left) and the future climate (right). 195

Figure 7.6: Relation between damage ratio and the maximum wind speed $y_{NHM,MAX}$ computed with the JMA-NHM at different locations and the estimated damage ratio model. 198

Figure 7.7: Cumulative annual average numbers of typhoons in the AGCM simulation and the Monte Carlo simulation as a function of the central pressure at latitudes 30No (top) and 35N o (bottom) under the current climate (left) and the projected future climate (right). 202

Figure 7.8: Exceedance probabilities of the annual maximum wind speeds under the current and the projected future climates at Tokyo (left) and Fukuoka (right). 203

Figure 7.9: Comparisons of 50-year return period wind speeds (left) and medians of the annual maximum wind speeds (right) at 2249 locations in Japan under the current and the projected future climates. 204

Figure 7.10: Change of the typhoon wind risks under the current and projected future climates. 205

Figure 7.11: Geographical distribution of the change of the residential building risks. 205

Figure 7.12: Cumulative annual average numbers of typhoons in the JMA Best track data in the period of 1979-2003 and the AGCM simulation for the current climate as a function of the central pressure at latitudes 20N o (left) and 35N o (right)...................... 207

Figure 8.1: Interface for the option "conditional simulations"................ 212

Figure 8.2: Summary of the analysis results for the case of the analysis with the option "conditional simulation". 213

Figure 8.3: Probabilistic model representation. 218

Figure 8.4: Illustration of the transition of the typhoon and the location of the platform. 226

Figure 8.5: Event/decision tree of the decision problem considered in the example. 226

Figure 10.1: Cumulative distributions of the translation speed and direction and central pressure of typhoons in July. 251

Figure 10.2: Cumulative distributions of the translation speed and direction and central pressure of typhoons in August. 252

Figure 10.3: Cumulative distributions of the translation speed and direction and central pressure of typhoons in September. 253

Figure 10.4: Visualization of historical typhoon tracks and wind fields. 259

List of Tables

Table 2-1: Summary of utilized datasets. 40

Table 2-2: Categories of the land use data and the according roughness length. 71

Table 3-1: Comparison of 100-year and 500-year wind speeds. 91

Table 4-1: Summary of the alternative models. The differences relative to the reference model 0 are highlighted. 108

Table 5-1: Parameters of the prior distribution of a and b. 128

Table 6-1: Statistics assessed in the software tool *TRAST*. 162

Table 8-1: Conditions and associated losses postulated in the consequence model. 224

Table 8-2: Assumed initial conditions. 225

Table 10-1: Parameters determining Eg (escarpments) 254

Table 10-2: Parameters determining Eg (ridge-shaped topography) 255

i want morebooks!

Buy your books fast and straightforward online - at one of world's fastest growing online book stores! Environmentally sound due to Print-on-Demand technologies.

Buy your books online at
www.get-morebooks.com

Kaufen Sie Ihre Bücher schnell und unkompliziert online – auf einer der am schnellsten wachsenden Buchhandelsplattformen weltweit! Dank Print-On-Demand umwelt- und ressourcenschonend produziert.

Bücher schneller online kaufen
www.morebooks.de

VDM Verlagsservicegesellschaft mbH
Heinrich-Böcking-Str. 6-8
D - 66121 Saarbrücken

Telefon: +49 681 3720 174
Telefax: +49 681 3720 1749

info@vdm-vsg.de
www.vdm-vsg.de

Printed by Books on Demand GmbH, Norderstedt / Germany